LONDON MATHEMATICAL SOCIETY STUDENT TEXTS

Managing editor: Dr C.M. Series, Mathematics Institute
University of Warwick, Coventry CV4 7AL, United Kingdom

1 Introduction to combinators and λ-calculus, J.R. HINDLEY & J.P. SELDIN
2 Building models by games, WILFRID HODGES
3 Local fields, J.W.S. CASSELS
4 An introduction to twistor theory: Second edition, S.A. HUGGETT & K.P. TOD
5 Introduction to general relativity, L.P. HUGHSTON & K.P. TOD
6 Lectures on stochastic analysis: diffusion theory, DANIEL W. STROOCK
7 The theory of evolution and dynamical systems, J. HOFBAUER & K. SIGMUND
8 Summing and nuclear norms in Banach space theory, G.J.O. JAMESON
9 Automorphisms of surfaces after Nielsen and Thurston, A. CASSON & S. BLEILER
10 Nonstandard analysis and its applications, N. CUTLAND (ed)
11 Spacetime and singularities, G. NABER
12 Undergraduate algebraic geometry, MILES REID
13 An introduction to Hankel operators, J.R. PARTINGTON
14 Combinatorial group theory: a topological approach, DANIEL E. COHEN
15 Presentations of groups, D.L. JOHNSON
16 An introduction to noncommutative Noetherian rings, K.R. GOODEARL & R.B. WARFIELD, JR.
17 Aspects of quantum field theory in curved spacetime, S.A. FULLING
18 Braids and coverings: selected topics, VAGN LUNDSGAARD HANSEN
19 Steps in commutative algebra, R.Y. SHARP
20 Communication theory, C.M. GOLDIE & R.G.E. PINCH
21 Representations of finite groups of Lie type, FRANÇOIS DIGNE & JEAN MICHEL
22 Designs, graphs, codes, and their links, P.J. CAMERON & J.H. VAN LINT
23 Complex algebraic curves, FRANCES KIRWAN
24 Lectures on elliptic curves, J.W.S. CASSELS
25 Hyperbolic geometry, BIRGER IVERSEN
26 An introduction to the theory of L-functions and Eisenstein series, H. HIDA
27 Hilbert Space: compact operators and the trace theorem, J.R. RETHERFORD
28 Potential theory in the complex plane, T. RANSFORD
29 Undergraduate commutative algebra, M. REID
32 Lectures on Lie Groups and Lie Algebras, R. CARTER, G. SEGAL & I. MACDONALD
33 A primer of algebraic *D*-modules, S.C. COUTINHO

London Mathematical Society Student Texts 22

Designs, Graphs, Codes and their Links

P. J. Cameron
Professor of Mathematics, Queen Mary and Westfield College, London

and

J. H. van Lint
Professor of Mathematics, Eindhoven University of Technology

Published by the Press Syndicate of the University of Cambridge
The Pitt Building, Trumpington Street, Cambridge CB2 1RP
40 West 20th Street, New York, NY 10011-4211, USA
10 Stamford Road, Oakleigh, Melbourne 3166, Australia

First published 1991
Reprinted 1996

Library of Congress cataloging in publication data available

A catalogue record for this book is available from the British Library

ISBN 0 521 41325 7 hardback
ISBN 0 521 42385 6 paperback

Transferred to digital reprinting 2000

Contents

naller additions.

We are grateful to large numbers of students and colleagues for bringing errors ıd difficulties in the previous version to our attention, or for commenting on draft ıapters of this book. Especially, Rosemary Bailey, Frank De Clerck, Jon Hall, and ef Thas have given us great help.

We have used electronic mail to send the manuscript between one another, plain EX for the typesetting, and the laser printer in the School of Mathematical Sciences ; Queen Mary and Westfield College, London, to produce camera-ready copy.

As we said, this is not a textbook on graphs, codes, or designs — our treatment f these is mostly limited to the particular interconnections we discuss — and our eaders may require more extended treatments. We refer them to Hughes and Piper .985) for designs, Beineke and Wilson (1978), (1983) for graphs, and van Lint (1982) r codes. Though we give many references, it is not compulsory to read all these!

We assume a background of undergraduate algebra. The following list of results ıcludes most of what is needed. These results can be found in any good algebra extbook. We give references to the two-volume *Algebra*, by P. M. Cohn (1974), 1977).

- Finite fields have prime power order. For each $q = p^d$ (with p prime and $d > 0$), there is a unique field of order q, up to isomorphism (the Galois field). Its multiplicative group is cyclic of order $q - 1$, and its automorphism group is cyclic of order d. (Cohn (1977), p. 195.)
- For any real symmetric matrix A, there is an orthonormal basis of $\mathbf{R}^n$ consisting of eigenvectors of A. Equivalently, there is an orthogonal matrix P such that $PAP^\top$ is diagonal. (Cohn (1974), p. 203.)
- For any $m \times n$ integer matrix A, there are integer matrices P and Q with determinant 1 such that $PAQ = \begin{pmatrix} D & O \\ O & O \end{pmatrix}$, where $D = \mathrm{diag}(d_1, \ldots, d_r)$, the d_i being non-zero integers satisfying $d_1 | d_2 | \ldots | d_r$. (This is the Smith normal form of A; the numbers d_i are the elementary divisors of A.) (Cohn (1974), p. 279.)
- The polynomial ring over a field, or any quotient of this ring, is a principal ideal domain. (Cohn (1974), pp. 134, 276.)
- If A is a finite abelian group, then the group of characters of A (homomorphisms to the multiplicative group of non-zero complex numbers) is isomorphic to A. (Cohn (1974), p. 243; (1977), p. 163.) (In fact, we only use this for groups A which are given as direct sums of cyclic groups, so the 'Fundamental Theorem of Abelian Groups' is not required.)
- We also assume familiarity with the concepts of a group and the action of a group on a set, and a few specific groups (Cohn (1974), p. 47.)

n the other hand, we have made no assumptions about prerequisites in discrete ıathematics. Usually, what we need is developed in the text; but we have included ı account of the Principle of Inclusion and Exclusion as an appendix to Chapter 1. See Hall (1986) for further discussion.)

Preface

The three subjects of this book all began life in the provinces of
mathematics. Design theory originated in statistics (its name reflects its in
experimental design); codes in information transmission; and graphs in the
of networks of a very general kind (in the first instance, the bridges of K
All three have since become part of mainstream discrete mathematics.

We have not tried to write a textbook on three individual topics. In
goal is more limited: we want to explore some of the ways in which the t
have interacted with each other, with results and methods from one area be
in another. Indeed, we believe that discrete mathematics is better defi
methods than by its subject-matter, and our approach reflects this.

The book has its origins in the notes of two series of lectures giv
authors at Westfield College, London, at the invitation of Dan Hughes. Th
at those lectures consisted of design theorists, and our job was to show
graphs and codes could be useful to them. The notes subsequently appea
London Mathematical Society Lecture Note Series in 1975, and in a co
revised form in 1980. We tried then to make the notes accessible to a wide
by adding an introductory chapter on design theory.

In the intervening decade, we have become aware that a number of stu
the book as a textbook. Their task was not made easier by the 'research r
in which many assertions are left without proof. Accordingly, when Dav
approached us about a new revision, we decided to re-write the book c
turning it into a textbook. We have expanded considerably the chapters
theory, strongly regular graphs, and codes; we have, wherever possible
proofs of our assertions, and avoided words like 'clearly ...'; and we hav
number of exercises, with hints where appropriate.

In addition, we have brought the material up-to-date, with a numb
topics (including graphs with least eigenvalue -2 and their connection
systems, strongly regular graphs with strongly regular subconstituents, and
treatments of two-graphs, partial geometries, Preparata and Kerdock co
weight projective codes, P- and Q-polynomial association schemes, etc.),

Notation and terminology

Our terminology is mostly standard, with a few exceptions. In particular, we use the term 'square 2-design' for what is usually called 'symmetric design', '$(0,k)$-set' for 'maximal k-arc', and 'ball' for 'sphere'. Our reasons for these decisions are explained in the text.

In many instances, the symbols commonly used for parameters of designs, graphs, geometries, etc. conflict with one another. There is no simple solution to this, and we have been deliberately inconsistent, in the interests of clarity. The reader's attention is drawn to this whenever it occurs.

We use F_q to denote the Galois field with q elements, where q is a prime power. (This field is often called GF(q).) Vectors, over a finite field or the real numbers, are written in bold face.

If $\mathbf{u}$ and $\mathbf{v}$ are vectors in some vector space, then
$\langle \mathbf{u}, \mathbf{v} \rangle$ denotes their dot product;
$[\mathbf{u}, \mathbf{v}]$ denotes the subspace they span;
$(\mathbf{u}, \mathbf{v})$ denotes simply the ordered pair.
The square brackets for span are also used for arbitrary sets of vectors.

An n-tuple of scalars in round brackets denotes a row vector, in the usual way. Matrices are denoted by capital letters, and the transpose of A is written A^{T}. This notation is also extended to denote 'duals' of other algebraic or combinatorial objects (for example, designs), as explained in the text.

1. Design theory

In this chapter, we describe some concepts and results from design theory, and construct some important designs.

(1.1) DEFINITION. A *t-design* with parameters (v, k, λ) (or a t-(v, k, λ) design) is a pair $\mathcal{D} = (X, \mathcal{B})$, where X is a set of 'points' of cardinality v, and $\mathcal{B}$ a collection of k-element subsets of X called 'blocks', with the property that any t points are contained in precisely λ blocks.

Various conditions are usually appended to the definition to exclude degenerate cases. We assume that X and $\mathcal{B}$ are non-empty, and that $v \geq k \geq t$ (so that $\lambda > 0$). A t-design with $\lambda = 1$ is called a *Steiner system.* (The notation $S(t, k, v)$ is also used for Steiner systems, and is sometimes extended to $S_\lambda(t, k, v)$ for arbitrary designs.) Alternatively, a t-design can be defined to consist of a set X of points and a set $\mathcal{B}$ of blocks, with a relation of 'incidence' between points and blocks, satisfying the appropriate conditions, including the assertion that k distinct points are incident with at most one block.

Sometimes a t-design is redefined so as to allow 'repeated blocks', that is, $\mathcal{B}$ is a family rather than a set of sets, and the same k-element set of points may occur more than once as a block. (This is more natural if we adopt the 'incidence relation' definition; simply omit the condition that k points are incident with at most one block.) In this book, we normally do not allow repeated blocks; where they are permitted, we will say so. Following Hughes and Piper (1985), we often use the term *t-structure* to signify that repeated blocks are permitted.

It is worth a short digression to explain the force of the 'no repeated blocks' assumption.

The question of the existence of designs with specified parameters has very different answers depending on whether we allow repeated blocks or not. If $\mathcal{B}$ is the set of all k-subsets of X, then $(X, \mathcal{B})$ is trivially a t-design for any $t \leq k$. If repeated blocks are permitted, we have the following easy result.

(1.2) Proposition. *Suppose that $t < k < v - t$. Then there is a t-(v, k, λ) structure for some λ, in which not every k-set of points is incident with a block.*

PROOF. Let M be the $\binom{v}{k} \times \binom{v}{t}$ matrix whose rows and columns are indexed by the k-subsets and t-subsets of X respectively, in which the entry indexed by (K, T) is 1 if $T \subset K$, 0 otherwise. By hypothesis, $\binom{v}{t} < \binom{v}{k}$; so the rows of M are linearly dependent over $\mathbf{Q}$. So there is a vector $\mathbf{v} \in \mathbf{Q}^{\binom{v}{k}}$ such that $\mathbf{v}M = \mathbf{0}$. By multiplying by the least common multiple of the denominators of the entries of $\mathbf{v}$, we may assume that these entries are integers. Let $-m$ be the smallest entry of $\mathbf{v}$. If $\mathbf{1}$ denotes the all-1 vector, then $\mathbf{w} = \mathbf{v} + m\mathbf{1}$ is a non-negative integer vector with at least one component 0; and

$$\mathbf{w}M = (\mathbf{v} + m\mathbf{1})M = m\mathbf{1}M = m\binom{v-t}{k-t}\mathbf{1}.$$

The interpretation of this matrix equation is that, if $\mathcal{B}$ is the family of k-subsets of X in which the set K is repeated $\mathbf{w}_K$ times, then any t-set lies in $\lambda = m\binom{v-t}{k-t}$ members of $\mathcal{B}$ (counted with multiplicity). So $(X, \mathcal{B})$ is the required structure. □

It is possible to place some conditions on the value of λ as well: see Wilson (1973).

The situation is quite different, however, if repeated blocks are forbidden. The t-design condition becomes stronger as t increases (see (1.5)). A couple of 5-designs have been known for most of this century; but, at the time the previous version of this book was written, no non-trivial 6-design was known. Since then, first Magliveras and Leavitt (1983), and later others, found some particular 6-designs; then Teirlinck (1987), (1989) spectacularly resolved the existence question by proving the following result.

(1.3) Theorem. *Given t, let*

$$\mu = \prod_{i=1}^{t}\left(\mathrm{lcm}\{\binom{i}{n} : n = 1, \ldots, i\} \cdot \mathrm{lcm}\{1, \ldots, i+1\}\right).$$

Then, for any $v \equiv t \pmod{\mu}$, the set of all $(t+1)$-subsets of a v-set X can be partitioned into t-$(v, t+1, \mu)$ designs. In particular, a non-trivial t-$(v, t+1, \lambda)$ design exists whenever $v \equiv t \pmod{\mu}$, $\lambda \equiv 0 \pmod{\mu}$, and $v > \lambda + t$. □

However, necessary and sufficient conditions for the existence of t-designs are far from being known, even asymptotically. In particular, there are still no known examples of Steiner systems with $t \geq 6$, and only finitely many with $t \geq 4$. There are 'classical' 5-(24, 8, 1) and 5-(12, 6, 1) designs constructed by Skolem (1931), Witt (1938a,b); the other known Steiner systems with $t = 5$ can all be found in Denniston (1976) and Mills (1978). The existence of Steiner systems with large t is possibly the most important open problem in design theory.

An *isomorphism* from $(X, \mathcal{B})$ to $(X', \mathcal{B}')$ is a one-to-one map f from X to X' which carries each set in $\mathcal{B}$ to a set in $\mathcal{B}'$, and such that each set in $\mathcal{B}'$ occurs as the image of a unique set in $\mathcal{B}$. Isomorphic designs may be regarded as being structurally 'the same'. We will encounter some very nice situations in which a design $\mathcal{D}$ is 'characterized' by its parameters, in the sense that any design with the same parameters is isomorphic to $\mathcal{D}$.

(If repeated blocks are allowed, this definition of isomorphism would not suffice; we should have to define an isomorphism to be a pair of bijections, from X to X' and from $\mathcal{B}$ to $\mathcal{B}'$, preserving incidence and non-incidence. We will ignore this complication.)

The set of *automorphisms* of a design (that is, isomorphisms from the design to itself) forms a group. Moreover, this automorphism group acts in a natural way as a permutation group on the points of the design, or on its blocks. Group theory provides very powerful tools for studying permutation groups (see Wielandt (1964), Cameron (1981), for example). For the most part, we will not consider these, except for occasionally using a group in one of our constructions.

We now derive some simple necessary conditions for the existence of a design.

(1.4) Proposition. *Let $\lambda(S)$ be the number of blocks containing a given set S of s points in a t-(v, k, λ) design, where $0 \leq s \leq t$. Then*

$$\lambda(S)\binom{k-s}{t-s} = \lambda\binom{v-s}{t-s}.$$

PROOF. Count the number of choices of a block B containing S and $t - s$ further points of B, to obtain the result. □

Note that $\lambda(S)$ depends only on the cardinality s of S; so we will write it as λ_s. It satisfies

$$\lambda_s\binom{k-s}{t-s} = \lambda\binom{v-s}{t-s}. \tag{1.5}$$

From these remarks, two corollaries follow:

(1.6) Corollary. *A t-design is also a s-design for $0 \leq s \leq t$.* □

(1.7) Corollary. *If a t-(v, k, λ) design exists, then*

$$\binom{k-s}{t-s} \text{ divides } \binom{v-s}{t-s}\lambda,$$

for $s = 0, \ldots, t-1$. □

It is virtually a universal convention in design theory to denote λ_0 (the total number of blocks) by b, and (if $t \geq 1$) to denote λ_1 (the number of blocks containing a point) by r. Since any t-design for $t \geq 1$ can be regarded as a 1-design by (1.4), we can apply (1.5) to obtain:

(1.8)
$$bk = vr.$$

A 2-design is often called a *block design* or simply a *design.* In the literature the term 'balanced incomplete-block design' is used, abbreviated to BIBD. (Balance refers to the 2-design condition, and incompleteness to the fact that $k < v$.) An alternative term is 'pairwise balanced design', though such designs need not have constant block size. In a 2-design, we have:

(1.9)
$$r(k-1) = (v-1)\lambda.$$

(1.10) DEFINITION. An *incidence matrix* of a design is a matrix M whose rows and columns are indexed by the blocks and points of the design respectively, the entry indexed by (B, p) being 1 if $p \in B$, 0 otherwise.

The incidence matrix depends on the ordering chosen for points and blocks. (The reader is warned that a different convention is often used, for example in the books by Dembowski (1968) and Hall (1986), with the result that our incidence matrices are the transposes of the ones appearing in those books. The present convention is adopted because we shall want to regard the characteristic functions of blocks, or rows of M, as row vectors, and consider the subspace they span.)

The conditions that any block contains k points, any point lies in r blocks, and any pair of points lies in λ blocks, can be expressed in terms of M:

(1.11)
$$MJ = kJ,$$
$$JM = rJ,$$
$$M^{\top}M = (r-\lambda)I + \lambda J.$$

(Here, as throughout this book, I is an identity matrix, and J a matrix with every entry 1, of the appropriate size.)

(1.12) Lemma. *If I and J are the identity and all-1 matrices of order n, then*

$$\det(xI + yJ) = (x + yn)x^{n-1}.$$

PROOF. $xI + yJ$ is symmetric, and so has an orthonormal basis of eigenvectors; its determinant is the product of its eigenvalues. Now the all-1 vector $\mathbf{1}$ is an eigenvector with eigenvalue $x + yn$. Any other eigenvector $\mathbf{v}$ is orthogonal to $\mathbf{1}$, and so $\mathbf{v}J = \mathbf{0}$, and $\mathbf{v}(xI + yJ) = x\mathbf{v}$. So x is an eigenvalue with multiplicity $n - 1$. □

Suppose that $\lambda > 0$ and $k < v$. By (1.9), $\lambda(v-k) = (r-\lambda)(k-1)$, and so $r - \lambda > 0$. Now by (1.11) and (1.12),

$$\det(M^{\top}M) = \det\left((r-\lambda)I + \lambda J\right) = rk(r-\lambda)^{v-1}, \tag{1.13}$$

and so $M^{\top}M$ is non-singular. *Fisher's inequality* follows:

(1.14) Theorem. *In a 2-design with $k < v$, we have $b \geq v$.* □

Furthermore, if $b = v$, then $r = k$, and so $MJ = JM$; thus M commutes with $(r-\lambda)I + \lambda J$, and so also with $((r-\lambda)I+\lambda J)M^{-1} = M^{\top}$. So $MM^{\top} = (r-\lambda)I + \lambda J$, from which it follows that any two blocks have exactly λ points in common.

(1.15) Theorem. *In a 2-design with $k < v$, the following conditions are equivalent:*
(a) $b = v$;
(b) $r = k$;
(c) any two blocks have λ common points;
(d) any two blocks have a constant number of common points.

PROOF. We have seen the implications (a) ⇔ (b) and (b) ⇒ (c), while (c) ⇒ (d) is trivial.

For the last step, we need the concept of the *dual* of a design $\mathcal{D} = (X, \mathcal{B})$. This is the design $\mathcal{D}^{\top} = (X^{\top}, \mathcal{B}^{\top})$, where $X^{\top} = \mathcal{B}$, and $\mathcal{B}^{\top} = \{\beta_x : x \in X\}$, where

$$\beta_x = \{B \in \mathcal{B} : x \in B\}.$$

(If we had used the 'incidence relation' definition of a design, we could simply say $\mathcal{B}^{\top} = X$, and the incidence relation in $\mathcal{D}^{\top}$ is the converse of that in $\mathcal{D}$.)

The dual of a 1-design is a 1-design. and is a 2-design if and only if (1.15)(d) holds. Thus, if a 2-design satisfies (1.15)(d), then $b \geq v$ (by (1.14)) and $v \geq b$ (applying (1.14) to the dual design); so $b = v$. □

(1.16) REMARK. The notation $\mathcal{D}^{\top}$ is intended as an *aide-mémoire*, since the incidence matrix of $\mathcal{D}^{\top}$ is the transpose of that of $\mathcal{D}$.

(1.17 DEFINITION. A 2-design is called *square* if it satisfies the equivalent conditions of (1.15).

This terminology is not standard. Dembowski, in his influential book (1968), used the term 'projective', for reasons which will appear shortly. But the most common term is 'symmetric'. This is unsatisfactory, since it suggests a stronger condition, viz. isomorphism of the design with its dual, which doesn't hold in all square 2-designs. We now explore this concept.

(Note: The term 'square' has been applied to arbitrary designs or incidence structures which have equally many points and blocks.)

(1.18) DEFINITION. A *duality* of a design $\mathcal{D}$ is an isomorphism from $\mathcal{D}$ to its dual. It can be described as a pair of bijections $\sigma : X \to \mathcal{B}$ and $\tau : \mathcal{B} \to X$ such that

$$x \in B \text{ if and only if } B^\tau \in x^\sigma.$$

The result of applying the duality twice is the pair of maps $\sigma\tau : X \to X$ and $\tau\sigma : \mathcal{B} \to \mathcal{B}$, which give an automorphism of the design. The duality is called a *polarity* if this automorphism is trivial, that is, τ is the inverse of σ; in this case, the polarity is determined by the single map σ, which satisfies

$$x \in y^\sigma \text{ if and only if } y \in x^\sigma.$$

(1.19) Proposition. *A design admits a polarity if and only if it has a symmetric incidence matrix (relative to some ordering of points and blocks).*

PROOF. If σ is a polarity, and $X = \{x_1, \ldots, x_v\}$, then relative to this ordering of points and the ordering $\{x_1^\sigma, \ldots, x_v^\sigma\}$ for blocks, the incidence matrix is symmetric. The converse is similar. □

We return to polarities in the next chapter.

(1.14) and (1.15) follow from a more general result, which we will need in Chapter 7. It also introduces a very useful technique, the 'variance trick'.

(1.20) Theorem. *Let B be a block of a 2-(v, k, λ) design. Then the number of blocks not disjoint from B is at least $k(r-1)^2/((k-1)(\lambda-1)+(r-1))$. Equality holds if and only if blocks which are not disjoint from B meet it in a constant number of points. If this occurs, then the constant number is $1+(k-1)(\lambda-1)/(r-1)$.*

PROOF. Let d be the number of blocks which are distinct from but not disjoint from B; suppose that n_i of these blocks meet B in i points. Count in two ways the number of choices of j points in B and a block (different from B) containing them, for $j = 0, 1, 2$. We obtain the following equations, where the summation is over i running from 1 to k.

$$\sum n_i = d,$$

$$\sum i n_i = k(r-1),$$

$$\sum i(i-1) n_i = k(k-1)(\lambda-1).$$

So

$$\sum (i-x)^2 n_i = dx^2 \; - \; 2k(r-1)x \; + \; k((k-1)(\lambda-1)+(r-1)).$$

This quadratic form in x must be positive semi-definite, proving the inequality. It vanishes only if $d = k(r-1)^2/((k-1)(\lambda-1)+(r-1))$, in which case $n_i = 0$ for all $i \neq 1+(k-1)(\lambda-1)/(r-1)$. □

Now Fisher's inequality follows from $b-1 \geq k(r-1)^2/((k-1)(\lambda-1)+(r-1))$, using (1.8) and (1.9) and a little calculation. (A hint for the calculation: express everything in terms of the parameters r, k, λ; after clearing the denominator, the difference between these two expressions is $(r-\lambda)^2(r-k)(k-1)$.) Also, if $b = v$, then $r = k$, and $1 + (k-1)(\lambda-1)/(r-1) = k$.

The *Bruck–Ryser–Chowla theorem* gives a further necessary condition on the existence of a square 2-(v, k, λ) design, beyond the equation $k(k-1) = (v-1)\lambda$ which follows from (1.9) and (1.15).

(1.21) Theorem. *Suppose that there exists a square 2-(v, k, λ) design. Set $n = k - \lambda$. Then*
(a) if v is even then k is a square;
(b) if v is odd, then the diophantine equation

$$z^2 = nx^2 + (-1)^{(v-1)/2}\lambda y^2$$

has a solution in integers x, y, z, not all zero.

PROOF. (a) is immediate from the fact that

$$\det(M)^2 = \det(M^\top M) = k^2 n^{v-1},$$

see (1.9) and (1.11). We do not offer a proof of (b). Several different proofs are available, of which the most familiar relies on Lagrange's results on sums of squares, and others use Hasse–Minkowski theory or coding theory. We refer to Hughes and Piper (1985). □

It is now known that the condition $k(k-1) = (v-1)\lambda$ and the Bruck–Ryser–Chowla theorem are not sufficient for the existence of a square 2-design. One single parameter set, viz. 2-(111, 11, 1), is known which satisfies these conditions where no design exists. (We have more to say about this case later.) For any given value greater than 1 of λ, only finitely many square 2-(v, k, λ) designs are known to exist.

We now look at a couple of special classes of square designs.

A (finite) *projective plane* of *order* n is a 2-$(n^2+n+1, n+1, 1)$ design. Projective planes are known to exist for all prime power orders, but no plane of non prime power order is known. The most familiar projective planes are the so-called *Desarguesian* planes. These are special cases of *projective geometries*, which we now define.

(1.22) EXAMPLE. Let $\mathsf{F} = \mathsf{F}_q$, and let V be a $(n+1)$-dimensional vector space over F. The *projective space* or *projective geometry* $\mathrm{PG}(n, q)$ consists of the set of all vector subspaces of V. It can be regarded as a partially ordered set (where the ordering is set-theoretic inclusion — it is in fact a lattice with respect to this ordering), or as an 'incidence structure' in which two subspaces are incident whenever one contains the other. An *i-flat* is a subspace of vector space dimension $i+1$; 0-flats, 1-flats, 2-flats and $(n-1)$-flats are called *points, lines, planes, hyperplanes* respectively.

It is clear that two subspaces are equal if and only if they contain the same points. So we can regard the set of points as basic, and identify any flat with the set of points it contains. The design theorist's interest in this procedure is that, for any fixed i with $1 \leq i \leq n-1$, the points and i-flats form a 2-design. In particular, the points and lines form a Steiner system (a 2-$((q^{n+1}-1)/(q-1), q+1, 1)$ design); while the points and hyperplanes form a square 2-design (which is a 2-$((q^{n+1}-1)/(q-1), (q^n-1)/(q-1), (q^{n-1}-1)/(q-1))$ design). We sometimes use the notation $\mathrm{PG}(n,q)$ to denote the point-hyperplane design.

The intersection of these two cases is the design of points and lines in $\mathrm{PG}(2,q)$, which is a 2-$(q^2+q+1, q+1, 1)$ design, that is, a projective plane of order q.

We now give some characterizations of projective spaces as designs. For undefined terms such as 'Desargues' Theorem', we refer to Hughes and Piper (1973).

(1.23) Theorem. *For a projective plane $\mathcal{D}$, the following conditions are equivalent:*
(a) $\mathcal{D}$ is the point-line design of $\mathrm{PG}(2,q)$ *for some prime power q;*
(b) $\mathcal{D}$ satisfies Desargues' Theorem;
(c) $\mathcal{D}$ satisfies Pappus' Theorem;
(d) $\mathrm{Aut}(\mathcal{D})$ *is 2-transitive on the points of $\mathcal{D}$.* □

(The last condition means that any two distinct points can be mapped to any other two distinct points by an automorphism of $\mathcal{D}$.)

(1.24) Theorem. *For a 2-$(v,k,1)$ design $\mathcal{D}$ with $v > k > 2$, which is not a projective plane, the following conditions are equivalent:*
(a) $\mathcal{D}$ is the point-line design of $\mathrm{PG}(n,q)$ *for some prime power q and some integer $n \geq 3$;*
(b) if a,b,c,d are four points such that the lines ab and cd are concurrent, then the lines ac and bd are concurrent. □

(The *line ab* here means the unique block containing a and b; two lines are *concurrent* if they meet in a point. This result is due to Veblen and Young (1916).)

Point-hyperplane designs of projective geometries may be recognized by the *Dembowski–Wagner theorem* (1960). In any 2-design, the *line* joining two distinct points p,q is defined to be the intersection of all blocks containing p and q. It is straightforward to show that two points lie on a unique line.

(1.25) Theorem. *Let $\mathcal{D}$ be a square 2-design with $\lambda > 1$. Then the following are equivalent:*
(a) $\mathcal{D}$ is a projective geometry;

(b) every line meets every block;
(c) the number of blocks containing three non-collinear points is constant. □

The Bruck–Ryser–Chowla theorem shows that if a projective plane of order $n \equiv 1$ or $2 \pmod 4$ exists, then n is the sum of two squares. (In this case, $v = n^2+n+1 \equiv 3 \pmod 4$, and so the diophantine equation is $y^2+z^2 = nx^2$. Standard reduction arguments, as in Hardy and Wright (1981), p. 301, show that, if this equation has a non-zero solution, then it has one with $x = 1$.) Thus there is no projective plane of order 6. As noted above, this theorem does not preclude the existence of a projective plane of order 10. The non-existence of such a plane was shown by several massive computations by Lam *et al.* (1983), (1986), (1989), using a coding-theoretic approach due to MacWilliams, Sloane and Thompson (1972). In Chapter 13, we describe the method used.

A *subplane* of a projective plane $(X, \mathcal{B})$ of order n consists of a proper subset X' of the point set X, and a subset $\mathcal{B}'$ of the line set $\mathcal{B}$, such that $(X', \mathcal{B}')$ is itself a projective plane (of order m, say). In this situation, $n \geq m^2$ holds (see Exercise 16); equality holds if and only if every line in $\mathcal{B}$ contains a point of X' (and dually). In the situation of equality, $(X', \mathcal{B}')$ is called a *Baer subplane* of $(X, \mathcal{B})$; and X' is a set of $m^2 + m + 1$ points which intersects every line in 1 or $m + 1$ points.

Another class of symmetric designs arises from Hadamard matrices, so-called because of their relationship to a theorem of Hadamard (1893).

(1.26) Theorem. *Let A be a $n \times n$ real matrix whose entries satisfy $|a_{ij}| \leq 1$ for all i, j. Then $|\det(A)| \leq n^{\frac{n}{2}}$. Equality holds if and only if all entries of A are ± 1 and $AA^{\top} = nI$.*

PROOF. $|\det(A)|$ is the volume of the n-dimensional parallelepiped spanned by the rows of A. By assumption, each row vector has Euclidean length at most $n^{\frac{1}{2}}$, with equality if and only if all its entries are ± 1. Also, the volume is at most the product of the edge lengths, with equality if and only if the edges are mutually perpendicular. □

(1.27) DEFINITION. A $n \times n$ real matrix H with entries ± 1 satisfying $HH^{\top} = nI$ is called a *Hadamard matrix* (or *H-matrix*, for short) of *order n*.]

Apart from trivial examples of orders 1 and 2, any Hadamard matrix has order divisible by 4, as we will see. It is conjectured that Hadamard matrices exist for all orders divisible by 4. The smallest multiple of 4 for which no Hadamard matrix is known is currently 428.

The defining property of a Hadamard matrix is unaltered if some rows or columns are multiplied by -1, or if rows or columns are permuted. We call two Hadamard matrices *equivalent* if one can be transformed into the other by such operations.

Any Hadamard matrix is equivalent (by sign changes alone!) to a *normalized* Hadamard matrix, in which the first row and column consist entirely of +1s. Let M be the $(n-1)\times(n-1)$ matrix obtained from a normalized Hadamard matrix by deleting the first row and column and replacing the -1s by 0s. Then M is the incidence matrix of a square $2\text{-}(v, \frac{1}{2}(v-1), \frac{1}{4}(v-3))$ design, where $v = n-1$. For any two rows of H agree in $\frac{1}{2}n$ positions, and so a row of M agrees with the all-1 vector in $\frac{1}{2}n - 1 = \frac{1}{2}(v-1)$ positions, i.e. has $\frac{1}{2}(v-1)$ entries 1. Moreover, suppose that two rows of M have x common ones. Then there are $\frac{1}{2}(v-1) - x$ positions where the entries in the two rows are 1 and 0, and the same number where they are 0 and 1; hence there are $x+1$ where both have the entry 0, and we conclude that $2x + 1 = \frac{1}{2}(v-1)$, or $x = \frac{1}{4}(v-3)$. Similar remarks apply to columns. So M is the incidence matrix of a design, as claimed. Note that $v \equiv 3 \pmod 4$, so n is divisible by 4.

Conversely, if M is the incidence matrix of a square $2\text{-}(4\lambda+3, 2\lambda+1, \lambda)$ design, then replacing the zeros in M by -1s and bordering M with a row and column of +1s gives a Hadamard matrix of order $4\lambda + 4$. Hence:

(1.28) Proposition. *There exists a Hadamard matrix of order $n > 2$ if and only if there exists a square 2-design with parameters $(n-1, \frac{1}{2}n - 1, \frac{1}{4}n - 1)$.* □

A square 2-design with these parameters is called a *Hadamard 2-design*.

(1.29) REMARK. Isomorphic Hadamard 2-designs come from equivalent H-matrices; but the converse is not true.

(1.30) EXAMPLE. A class of examples is due to Paley (1933). Let q be a prime power congruent to 3 mod 4, and let $\mathsf{F} = \mathsf{F}_q$, and Q the set of non-zero squares in F. Then $(\mathsf{F}, \{Q + x : x \in \mathsf{F}\})$ is a Hadamard 2-design. These designs are called *Paley designs*, and the Hadamard matrices obtained from them as in (1.24) are called *Paley H-matrices.* We reserve the term *Paley matrix* for a slightly different object, defined for any odd prime power q, namely the $q \times q$ matrix $P = (p_{ij})$, where

$$p_{ij} = \begin{cases} 0 & \text{if } i = j, \\ 1 & \text{if } i - j \text{ is a square in } \mathsf{F}_q, \\ -1 & \text{otherwise,} \end{cases}$$

where the indices are taken to be elements of F_q. If $q \equiv 3 \pmod 4$, the Paley H-matrix is obtained from P by putting -1 on the diagonal and bordering with a row and column of 1s.

(1.31) EXAMPLE. The point-hyperplane design of $\mathrm{PG}(n,2)$ is a Hadamard $2\text{-}(2^{n+1} - 1, 2^n - 1, 2^{n-1} - 1)$ design.The corresponding Hadamard matrix is called a *Sylvester H-matrix*. Sylvester H-matrices have a number of remarkable properties. For example, the character table of an abelian group of exponent 2 is a Sylvester H-matrix. See also Exercise 2.

We turn to some simple constructions of new designs from old ones.

(1.32) DEFINITION. Let $\mathcal{D} = (X, \mathcal{B})$ be a t-(v, k, λ) design, and p a point. The *derived design* $\mathcal{D}_p$ has point set $X \setminus \{p\}$ and block set $\{B \setminus \{p\} : B \in \mathcal{B}, p \in B\}$. It is a $(t-1)$-$(v-1, k-1, \lambda)$ design. Note that derived designs of $\mathcal{D}$ with respect to different points need not be isomorphic.A design $\mathcal{E}$ is called an *extension* of $\mathcal{D}$ if $\mathcal{E}$ has a point p such that $\mathcal{E}_p$ is isomorphic to $\mathcal{D}$; we call $\mathcal{D}$ *extendable* if it has an extension.

Since an extension of a t-(v, k, λ) design is a $(t+1)$-$(v+1, k+1, \lambda)$ design, (1.7) has the following consequence:

(1.33) Proposition. *If a t-(v, k, λ) design has an extension, then $k+1$ divides $b(v+1)$.* □

(1.34) Proposition. *The only extendable projective planes are those of orders 2 and 4.*

PROOF. By (1.25), if a plane of order n is extendable, then $n+2$ divides $(n^2+n+2)(n^2+n+1)$. By the Remainder Theorem, $n+2$ divides 12, and so $n = 2, 4$ or 10. (This much is due to Hughes (1961).) As we saw already, there is no plane of order 10. But, in fact, a relatively small subset of the computation establishing this fact shows that no such plane is extendable. We return to this later. □

Cameron (1973a) considered the more general question: which square designs are extendable?

(1.35) Theorem. *If a square 2-(v, k, λ) design $\mathcal{D}$ is extendable, then one of the following holds:*
(a) $\mathcal{D}$ is a Hadamard 2-design;
(b) $v = (\lambda+2)(\lambda^2+4\lambda+2)$, $k = \lambda^2+3\lambda+1$;
(c) $v = 495$, $k = 39$, $\lambda = 3$.

PROOF. It is convenient to re-define the symbols so that $\mathcal{E}$ is a 3-(v, k, λ) design which is an extension of a square design. First note that this condition is equivalent to the statement that any two blocks of $\mathcal{E}$ intersect in 0 or $\lambda+1$ points (this means that any two blocks of a derived design $\mathcal{E}_p$ meet in λ points).

Let B be a block of $\mathcal{E}$. If $p, q \notin B$, then there are $k\lambda/(\lambda+1)$ blocks containing p and q and meeting B in $\lambda+1$ points; and so there are $(k-\lambda-1)/(\lambda+1)$ blocks containing p and q and disjoint from B. This means that the incidence structure $\mathcal{E}^0$, whose points are the points outside B and whose blocks are the blocks disjoint from B, is a 2-$(v-k, k, (k-\lambda-1)/(\lambda+1))$ design. By (1.8) and (1.9), the number of blocks of $\mathcal{E}^0$ is

$$\frac{(v-k)(v-k-1)(k-\lambda-1)}{k(k-1)(\lambda+1).}$$

$\mathcal{E}^0$ may be degenerate, having just a single block; if this occurs, then $v = 2k$, $k = 2(\lambda+1)$, and $\mathcal{E}$ is an extension of a Hadamard 2-design.

Otherwise, we can apply Fisher's inequality (1.14) to $\mathcal{E}^0$, obtaining

$$(v-k-1)(k-\lambda-1) \geq k(k-1)(\lambda+1),$$
$$(k-1)^2(k-(\lambda+1)(\lambda+2)) \geq 0,$$

(on putting $(v-2)\lambda = (k-1)(k-2)$), and so $k \geq (\lambda+1)(\lambda+2)$.

However,

$$\begin{aligned} b &= v(v-1)/k \\ &= (k^2-3k+2\lambda+2)(k^2-3k+\lambda+2)/k\lambda^2; \end{aligned}$$

so k divides $2(\lambda+1)(\lambda+2)$. The same expression shows that, if $k = 2(\lambda+1)(\lambda+2)$, then λ divides 3, so $\lambda = 1$ or 3. If $\lambda = 1$ then $\mathcal{E}$ is a 3-(112, 12, 1) design, i.e. an extension of a projective plane of order 10, and as we have seen, such a design does not exist. If $\lambda = 3$, then we obtain case (c) of the theorem.

Otherwise, we have $k = (\lambda+1)(\lambda+2)$, giving case (b) of the theorem.

In the Hadamard case, we have the following result.

(1.36) Proposition. *Any Hadamard 2-design has a unique extension.*

PROOF. Let $\mathcal{E}$ be an extension of a Hadamard 2-design. The argument in (1.26) shows that the complement of any block of $\mathcal{E}$ is a block. So, if $\mathcal{D} = \mathcal{E}_p = (X, \mathcal{B})$, then $\mathcal{E}$ has point set $X \cup \{p\}$ and block set

$$\{B \cup \{p\}, X \setminus B : B \in \mathcal{B}\},$$

and $\mathcal{E}$ is uniquely determined by $\mathcal{D}$. In fact, this construction can be applied to any Hadamard 2-design, and produces a 3-design (see Exercise 5). □

An extension of a Hadamard 2-design, that is, a 3-$(4\lambda+4, 2\lambda+2, \lambda)$ design, is called a *Hadamard 3-design*. If H is a Hadamard matrix, there is a simple description of an associated Hadamard 3-design, as follows. Normalize H so that all elements in the first row are $+1$. Then, in any row other than the first, there are $\frac{1}{2}n$ entries $+1$ and $\frac{1}{2}n$ entries -1. In this way, a row determines two $\frac{1}{2}n$-subsets of $\{1, \ldots, n\}$. The sets defined in this way by all rows other than the first are the blocks of a Hadamard 3-design. Any Hadamard 3-design arises in this manner.

Apart from Hadamard designs, the only known extendable square 2-design is the projective plane of order 4. This is also the only square 2-design which can be extended more than once, that is, for which an extension is itself extendable (see Exercise 6). We will construct its extensions at the end of this chapter.

For the next case, the non-existence of a 3-(57, 12, 2) design has been shown by Bagchi (1988, 1991).

If $\mathcal{D} = (X, \mathcal{B})$ is a t-(v, k, λ) design, the *residual* $\mathcal{D}^p$ *of* $\mathcal{D}$ *with respect to the point* $p \in X$ has point set $X \setminus \{p\}$ and block set $\{B \in \mathcal{B} : p \notin B\}$. It is a $(t-1)$-$(v-1, k, \lambda_{t-1} - \lambda_t)$ design.

Let $\mathcal{D} = (X, \mathcal{B})$ be a t-(v, k, λ) design. The *complementary design* $\overline{\mathcal{D}}$ is obtained by replacing each block of $\mathcal{D}$ by its complement; that is, $\overline{\mathcal{D}} = (X, \overline{\mathcal{B}})$, where

$$\overline{\mathcal{B}} = \{X \setminus B : B \in \mathcal{B}\}.$$

If $x_1, \ldots, x_t \in X$, then λ_j blocks contain any given j of these points. The Principle of Inclusion and Exclusion (see (1.56)) enables us to count the number of blocks containing none of $x_1, \ldots, x_t$. (Let A_i be the set of blocks which contain x_j. With the notation of (1.56), $|A(J)| = \lambda_j$.) We find that

$$\overline{\lambda} = \sum_{s=0}^{t} (-1)^s \binom{t}{s} \lambda_s. \tag{1.37}$$

In particular, for $t = 2$, we have

$$\overline{\lambda} = b - 2r + \lambda. \tag{1.38}$$

So:

(1.39) Proposition. *The complement of a* t-(v, k, λ) *design is a* t-$(v, v-k, \overline{\lambda})$ *design, where* $\overline{\lambda}$ *is given by* (1.37). □

Note that the complement of a square design is square.

When can we produce a design from $\mathcal{D}$ by removing a block B? In order that all blocks of the resulting structure have the same cardinality, we require that $|B \cap B'|$ is constant for $B' \neq B$, that is, that $\mathcal{D}$ is a square design. If so, and if $\mathcal{D}$ has parameters 2-(v, k, λ), then set $\mathcal{D}^B = (X \setminus B, \{B' \setminus B : B' \in \mathcal{B}, B' \neq B\})$. Then $\mathcal{D}^B$ is called the *residual* of $\mathcal{D}$ *with respect to the block* B. It is a 2-$(v-k, k-\lambda, \lambda)$ design. The unqualified term 'residual' will always refer to such a 'block-residual' rather than the 'point-residual' defined earlier. Different residuals of the same design need not be isomorphic!

If we set $v^B = v - k$, $k^B = k - \lambda$, and $\lambda^B = \lambda$, then

$$\lambda^B v^B = k^B(k^B + \lambda^B - 1).$$

It is possible that $\mathcal{D}^B$ has repeated blocks. But if this occurs, then necessarily $k - \lambda \leq \lambda$, whence $v \leq 2k - 1$; then residuals of the complement of $\mathcal{D}$ will have no repeated blocks.

A 2-(v, k, λ) design whose parameters satisfy $\lambda v = k(k + \lambda - 1)$ is called *quasi-residual.* Thus, as expected, any residual design is quasi-residual. We discuss the converse below.

We consider first the case $\lambda = 1$. An *affine plane* of order n is a 2-$(n^2, n, 1)$ design. Thus, any residual of a projective plane is an affine plane of the same order.

Conversely, let $(X, \mathcal{B})$ be an affine plane of order n. Call the blocks *lines*, and call two lines *parallel* if they are equal or disjoint. Then *Playfair's Axiom* (a version of the Euclidean parallel postulate) holds: if L is a line and p a point, there is a unique line parallel to L which contains p. It follows that parallelism is an equivalence relation on the set of lines, and that the lines in any parallel class partition the set of points. Hence any parallel class has n lines, and there are $b/n = n + 1$ parallel classes.

Let Π be the set of parallel classes, and $X^* = X \cup \Pi$. For each $L \in \mathcal{B}$, let $L^* = L \cup \{\pi\}$, where $\pi \in \Pi$ is the parallel class containing L. Set

$$\mathcal{B}^* = \{L^* : L \in \mathcal{B}\} \cup \{\Pi\}.$$

Then $(X^*, \mathcal{B}^*)$ is a projective plane of order n, whose residual with respect to the line Π is $(X, \mathcal{B})$.

This process, with its roots in the Renaissance theory of perspective, is called 'adjoining a line at infinity'. We summarize the conclusion in two forms.

(1.40) Proposition. *(a) There exists a projective plane of order n if and only if there exists an affine plane of order n.*

(b) A quasi-residual 2-design with $\lambda = 1$ is residual.

Hall and Connor (1953) proved the statement analogous to (1.40)(b) with $\lambda = 2$. We will give a proof of the Hall–Connor theorem in Chapter 6.

Affine planes are closely connected with orthogonal Latin squares. A *Latin square* of order n is an $n \times n$ array with entries $1, \ldots, n$ having the property that each element of $\{1, \ldots, n\}$ occurs exactly once in each row or column. Two Latin squares A, B of order n are *orthogonal* if, for any $x, y \in \{1, \ldots, n\}$, there exists a unique position (i, j) such that $a_{ij} = x$ and $b_{ij} = y$.

(1.41) Proposition. *Let $f(n)$ be the maximum number of mutually orthogonal Latin squares of order n. Then $f(n) \leq n - 1$, with equality if and only if there exists an affine plane of order n.*

PROOF. We construct a 'geometry' from a set $A_1, \ldots, A_r$ of mutually orthogonal Latin squares as follows. Take $X = \{1, \ldots, n\} \times \{1, \ldots, n\}$. The *lines* are of three kinds:

$$H_i = \{(x, i) : 1 \le x \le n\} \text{ for } 1 \le i \le n;$$
$$V_i = \{(i, y) : 1 \le y \le n\} \text{ for } 1 \le i \le n;$$
$$L_{ij} = \{(x, y) : (A_j)_{xy} = i\} \text{ for } 1 \le i \le n, 1 \le j \le r.$$

Each line has cardinality n. Any point of X lies on $r + 2$ lines (one H_i, one V_i, and one L_{ij} for each value of j), and two points lie on at most one line. The lines through a given point contain $(r+2)(n-1)$ further points, so $(r+2)(n-1) \le n^2 - 1$, or $r \le n - 1$. Equality holds if and only if any two points lie on a unique line, in which case the geometry is an affine plane.

Conversely, suppose that an affine plane of order n exists. Select two parallel classes, and number their lines H_i and V_i $(1 \le i \le n)$. The unique point in the intersection of V_i and H_j can be given 'coordinates' (i, j). Now, for each of the $n - 1$ further parallel classes, number the lines from 1 to n, and define an array $A = (a_{ij})$, where $a_{ij} = k$ if and only if the point (i, j) lies on the kth line of the class. These $n - 1$ square arrays are Latin squares, and are pairwise orthogonal. □

It is known that $f(n) \to \infty$ as $n \to \infty$ (Chowla, Erdős and Straus (1960)). For more on this function, see Wilson (1974a). The geometries constructed in the above proof are called *nets*, and are particular examples of *partial geometries*, which we discuss in Chapter 7.

Which affine planes are extendable? The necessary condition (1.33) for extendability is always satisfied. An extension of an affine plane, that is, a 3-$(n^2+1, n+1, 1)$ design, is called an *inversive plane*, or *Möbius plane*, of order n.

(1.42) EXAMPLE. Inversive planes of all prime power orders are known. For the simplest examples, let $X = \{\infty\} \cup \mathbb{F}_{q^2}$, and let $\mathcal{B}$ be the set of images of $B = \{\infty\} \cup \mathbb{F}_q$ under the *linear fractional group*

$$\mathrm{PGL}(2, q^2) = \left\{ z \mapsto \frac{az + b}{cz + d} : a, b, c, d \in \mathbb{F}_{q^2}, ad - bc \neq 0 \right\}.$$

There is a more geometric description. An *ovoid* in $\mathrm{PG}(3, q)$ is a set of $q^2 + 1$ points, no three collinear. It can be shown than any *plane* (that is, hyperplane) of $\mathrm{PG}(3, q)$ meets an ovoid O in either 1 or $q + 1$ points. Thus the plane sections of size $q + 1$ of O are the blocks of an inversive plane of order q. Any inversive plane arising in this way is called *egglike*. All known inversive planes are egglike.

An example of an ovoid is the *elliptic quadric*, the set of zeros of the quadratic form

$$x_1 x_2 + f(x_3, x_4),$$

where f is an irreducible quadratic form in two variables over $\mathbb{F}_q$. The resulting egglike inversive plane is the same as the one constructed above using $\mathrm{PGL}(2, q^2)$. (Note: When we refer to the set of points of projective space satisfying some equation, as here, we mean the set of 1-dimensional subspaces spanned by vectors satisfying the equation.)

If q is an odd power of 2, further examples are known, the *Suzuki–Tits ovoids*, see Lüneburg (1965).

The following result, due to Barlotti (1955) and Dembowski (1964b), summarizes our knowledge.

(1.43) Theorem. *(a) If q is odd, then any ovoid is projectively equivalent to the elliptic quadric; so there is a unique egglike inversive plane of order q.*

(b) If q is even, then any inversive plane of order q is egglike. □

REMARK. Very recently, J. A. Thas (unpublished) has shown that, if q is odd and $q \neq 11, 23$ or 59, then an inversive plane having at least one derived design isomorphic to $\mathrm{AG}(2, q)$ is isomorphic to the 'classical' model obtained from the elliptic quadric. Thus, for such q, $\mathrm{AG}(2, q)$ has a unique extension.

The property of parallelism in affine planes can be extended to a wider class of designs, the *affine designs*, as follows.

A *parallelism* of a design $\mathcal{D}$ (also called a *resolution* or 1-*factorization*) is a partition of the block set of $\mathcal{D}$ into classes $C_1, \ldots, C_r$ with the property that any point of $\mathcal{D}$ lies in a unique block of each class (so that the blocks in each class partition the point set). A design is called *resolvable* if it has a parallelism. A necessary condition for the existence of a parallelism is that k divides v. Kirkman's celebrated 'schoolgirls problem' (1847), (1850) asked for a resolvable 2-(15, 3, 1) design. Unlike the case of affine planes, there is no intrinsic definition of the parallelism; it may fail to exist, or there may be more than one.

The following result is due to Bose (1942).

(1.44) Theorem. *A resolvable 2-design satisfies $b \geq v + r - 1$ (equivalently $r \geq k + \lambda$). The bound is attained if and only if the cardinality of the intersection of two non-parallel blocks is constant.* □

We remark that the proof is similar to that of Fisher's inequality (1.14); indeed, there is a common proof of the two inequalities.

A 2-design attaining the bound in Bose's inequality is called an *affine design*, or *affine resolvable design*. In such a design, two blocks are parallel if and only if they are disjoint; so the parallelism has an intrinsic definition.

(1.45) EXAMPLE. Examples of affine designs are given by the *affine spaces*. Let V be an n-dimensional vector space over $\mathbb{F}_q$. The objects of the n-dimensional affine space are the cosets of vector subspaces of V. A coset of an i-dimensional subspace is called an *i-flat*, and (as for projective spaces) 0-, 1-, 2- and $(n-1)$-flats are called *points, lines, planes, hyperplanes* respectively. Note that points, or cosets of the zero subspace, are singletons, and can be identified with the vectors of V.

Now the points and hyperplanes of $\mathrm{AG}(n,q)$ form an affine 2-$\left(q^n, q^{n-1}, \frac{q^{n-1}-1}{q-1}\right)$ design. (Blocks are parallel if and only if they are cosets of the same subspace. If B, B' are non-parallel blocks, then $B \cap B'$ is a coset of an $(n-2)$-dimensional subspace, and so has cardinality q^{n-2}.) This design is the n-dimensional affine space over $\mathbb{F}_q$, which we also denote by $\mathrm{AG}(n,q)$.

Further examples are given by the Hadamard 3-designs, in which each parallel class consists of a block and its complement. $\mathrm{AG}(n,2)$ is the Hadamard 3-design arising from the Sylvester Hadamard matrix of order 2^n.

Dembowski (1964a) gave a common characterization of these two classes of affine designs, similar in spirit to the Dembowski–Wagner theorem (1960). There are also characterizations of the point-line designs of affine spaces, in the spirit of (1.23) and (1.24).

Summarizing, there is a close analogy between Bose's inequality, affine designs, and affine spaces on the one hand, and Fisher's inequality, square 2-designs, and projective spaces on the other. It was this analogy which led Dembowski to propose the term 'projective design' for what we have called 'square 2-design'.

Note that any affine design is quasi-residual. $\mathrm{AG}(n,q)$ is the residual of $\mathrm{PG}(n,q)$.

Our next topic in this chapter concerns ovals. Let $\mathcal{D}$ be a square 2-(v,k,λ) design. An *n-arc* is a set of n points of $\mathcal{D}$, no three of which are contained in a block. Given an n-arc S, a block B of $\mathcal{D}$ is called a *secant*, *tangent* or *passant* to S according as $|S \cap B| = 2, 1$ or 0.

(1.46) Proposition. *Any point of an n-arc in a square 2-(v,k,λ) design lies on $(n-1)\lambda$ secants and $k-(n-1)\lambda$ tangents. In particular, $n \le 1 + k/\lambda$.*

PROOF. Let S be an n-arc, and $p \in S$. Count pairs (q, B) where B is a secant containing p and q, for $q \in S \setminus \{p\}$. □

An n-arc A is called an *oval of type I* if each point of A lies on a unique tangent (that is, if $n = 1 + (k-1)/\lambda$), and an *oval of type II* if it has no tangents (that is, $n = 1 + k/\lambda$). Note that ovals can only exist if $\lambda | (k-1)$ or $\lambda | k$ respectively. In particular, the two types cannot coexist in the same design if $\lambda > 1$. Ovals of type

II meet every block in 0 or 2 points; they are sometimes called *hyperovals*, or *sets of class* $\{0, 2\}$.

(1.47) Proposition. *If a square 2-(v, k, λ) design has an oval of type II, then $k - \lambda$ is even.*

PROOF. Let S be the oval, and p a point outside S. The number of secants containing p is $\frac{1}{2}n\lambda = \frac{1}{2}(k + \lambda)$. □

(1.48) Proposition. *Let S be a Type I oval in a symmetric 2-(v, k, λ) design with $k - \lambda$ even. Then any point of the design lies on either one or all tangents to S.*

PROOF. Observe that k, λ and n are all odd, and so any point lies on at least one tangent to S. We apply a different version of the 'variance trick'. Let n_i be the number of points which lie on i tangents. Then

$$\sum n_i = v,$$
$$\sum i n_i = nk,$$
$$\sum i(i-1)n_i = n(n-1)\lambda.$$

Therefore

$$\sum (i-1)(i-n)n_i = 0,$$

whence every point lies on one or all the tangents. □

It follows from (1.48) that, in a projective plane of even order, all the tangents to a type I oval S pass through a point p, the *nucleus* or *knot* of S. Then $\{p\} \cup S$ is a Type II oval.

Let S be a type II oval in a 2-(v, k, λ) design $\mathcal{D}$. Then the structure whose points are the passants to S and whose blocks are the points outside S, with incidence the dual of incidence in $\mathcal{D}$, is a 2-$((k-2)(k-\lambda)/2\lambda, (k-\lambda)/2, \lambda)$ design.

Ovals are important in connection with extensions of projective planes. If $\mathcal{D}$ is a 3-design for which $\mathcal{D}_p$ is a projective plane, and B is any block not containing p, then B is a Type II oval in $\mathcal{D}_p$. (For, if q, r, s are collinear points of B, and L the line containing them, then these three points would lie on two blocks B and $L \cup \{p\}$ of $\mathcal{D}$.) In PG(2, 2), the Type II ovals are precisely the complements of lines, and all of them are blocks in the unique extension AG(3, 2). We will see how ovals occur in Lüneburg's construction of the extension of PG(2, 4). Also, the non-existence of a projective plane of order 10 containing an oval (and hence of an extendable plane of order 10) was shown by Lam *et al.* (1983) several years before they showed that no plane of order 10 exists.

Examples of Type I ovals in PG$(2, q)$ are provided by *conics*, the sets of zeros of non-singular quadratic forms. (We can take the quadratic form to be $xz - y^2$, where (x, y, z) are coordinates in the 3-dimensional vector space. Then the points of the conic are spanned by the vectors $(1, t, t^2)$ for $t \in \mathrm{F}_q$, together with $(0,0,1)$.) A celebrated theorem of finite geometry, *Segre's Theorem* (Segre (1954)), asserts that there are no others if q is odd.

(1.49) Theorem. *For q an odd prime power, any (Type I) oval in* $\mathrm{PG}(2, q)$ *is a conic.* □

The same conclusion holds for $q = 2$ or 4, but not for larger powers of 2. (A conic together with its nucleus is a Type II oval; removing a point other than the nucleus gives a Type I oval which is not a conic for $q \geq 8$, see Exercise 9. Moreover, for $q = 16$ or $q \geq 64$, there exist Type II ovals not of the form 'conic plus nucleus'.)

We turn next to an extension of Fisher's inequality for designs with larger values of t. This result is due to Ray-Chaudhuri and Wilson (1975); the case $t = 4$ had been proved earlier by Petrenjuk (1968).

(1.50) Theorem. *Let $\mathcal{D}$ be a t-(v, k, λ) design, with $t = 2s$ and $k \leq v - s$. Then $b \geq \binom{v}{s}$.*

PROOF. We use a modified incidence matrix M_s, whose rows are indexed by blocks and columns by s-sets of points, with (B, S) entry 1 if $S \subset B$, 0 otherwise. Then M_s is a $b \times \binom{v}{s}$ matrix, and it suffices to show that the rows of M_s span $\mathbf{R}^{\binom{v}{s}}$. Accordingly, let $\mathbf{r}_B$ be the row of M_s with label B, and let $\mathbf{e}_S$ be the vector with 1 in the position labelled S and 0 in all other positions (the unit basis vector corresponding to S).

For $0 \leq i \leq s$, set

$$\begin{aligned}
\mathbf{y}_i &= \sum_{|B \cap S| = i} \mathbf{r}_B \\
&= \sum_{j=0}^{i} \sum_{|S' \cap S| = j} \sum_{\substack{B \supseteq S' \\ |B \cap S| = i}} \mathbf{e}_{S'} \\
&= \sum_{j=0}^{i} \binom{s-j}{i-j} \nu_{2s-j, s+i-j} \left(\sum_{|S \cap S'| = j} \mathbf{e}_{S'} \right),
\end{aligned}$$

where $\nu_{m,n}$ denotes the number of blocks intersecting a given m-set M in a given n-subset N. (An argument using the Principle of Inclusion and Exclusion (1.56) shows that this number depends only on m and n if $n \leq m \leq t$.)

Set

$$\mathbf{x}_j = \sum_{|S' \cap S| = j} \mathbf{e}_{S'}.$$

Then we have a system of $s + 1$ linear equations for the $\mathbf{x}_j$ in terms of the $\mathbf{y}_i$. The coefficient matrix is triangular, and its diagonal entries $\nu_{2s-i,s}$ are non-zero (since

$s \leq k \leq v - s$). So the equations have a unique solution. In particular, $\mathbf{x}_s = \mathbf{e}_S$ is a linear combination of the $\mathbf{y}_i$, and so is in the row space of M_s, proving the theorem. □

Ray-Chaudhuri and Wilson also proved a 'dual' result.

(1.51) Theorem. *Let $s \leq k \leq v - s$, and let $\mathcal{B}$ be a collection of k-subsets of the v-set X having the property that, for $B, B' \in \mathcal{B}$, $B \neq B'$, the cardinality of $B \cap B'$ takes one of only s distinct values. Then $|\mathcal{B}| \leq \binom{v}{s}$.*

PROOF. Let M_i be the incidence matrix of blocks and i-sets, for $i \leq s$, as in (1.50); also, let N_{ij} be the incidence matrix of i-sets and j-sets, for $j \leq i \leq s$. Note that

$$M_i N_{ij} = \binom{k-j}{i-j} M_j.$$

Let $x_0 = k, x_1, \ldots, x_s$ be the cardinalities of block intersections. For $0 \leq i \leq s$, let A_i be the matrix with rows and columns indexed by blocks, with (B, B') entry equal to 1 if $|B \cap B'| = x_i$, 0 otherwise. (So $A_0 = I$.) The (B, B') entry of $M_i M_i^\top$ is the number of i-subsets of $B \cap B'$; so we have

$$M_i M_i^\top = \sum_{j=0}^{s} \binom{x_j}{i} A_j$$

for $i = 0, \ldots, s$. In this system of equations for the A_j in terms of the $M_i M_i^\top$, the coefficient matrix has (i, j) entry $\binom{x_j}{i}$, where i and j run from 0 to s. Elementary operations convert this into the Vandermonde matrix with (i, j) entry $(x_j)^i$, showing that it is invertible; so the equations can be solved for the A_j.

Let $\mathbf{v}$ be a vector in the null space of M_s. Because $M_s N_{si} = \binom{k-i}{s-i} M_i$, it follows that $\mathbf{v}$ lies in the null space of all M_i, and hence of all $M_i M_i^\top$, and hence of all A_j. But $A_0 = I$; so $\mathbf{v} = \mathbf{0}$. Thus the null space of M_s is zero, and $b \leq \binom{v}{s}$, as claimed. □

Remarkably, equality in these two results determines the same class of structures.

(1.52) Theorem. *Let $\mathcal{B}$ be a collection of k-subsets of the v-set X, where $2s \leq k \leq v - s$. Then any two of the following conditions imply the third:*

(a) $(X, \mathcal{B})$ is a $2s$-design;

(b) the cardinality of the intersection of two distinct sets in $\mathcal{B}$ takes just s distinct values;

(c) $|\mathcal{B}| = \binom{v}{s}$. □

(1.53) DEFINITION. A design attaining this bound is called a *tight $2s$-design*.

Obviously, any square 2-design is tight. Also, the trivial $2s$-design whose blocks are all the sets of size $k = v - s$ is tight. The combined efforts of Enomoto, Ito and Noda (1979) and Bremner (1979) have determined all tight 4-designs.

(1.54) Theorem. *Let $\mathcal{D}$ be a tight 4-(v, k, λ) design with $s > 1$ and $k < v - s$. Then $\mathcal{D}$ is the unique 4-$(23, 7, 1)$ design or its complement.* □

There are some non-existence results for larger values of t as well.

We will meet a further generalization of tight designs, in the concept of association schemes, in Chapter 17. It will turn out that there is a sense in which they are 'dual' to perfect codes.

The method of *intersection triangles* is useful for investigating t-designs, especially Steiner systems. Let $x_1, \ldots, x_l$ be points in a design such that, for $0 \le i \le l$, the number μ_i of blocks containing a given i points from this set depends only on i. The Principle of Inclusion and Exclusion shows that, for $0 \le j \le i \le l$, the number $\nu_{i,j}$ of blocks intersecting a given i of these points precisely in a given subset of size j depends only on i and j. The numbers $\nu_{i,j}$ can most easily be calculated by writing the $\mu_i = \nu_{i,i}$ on the right-hand border of a triangular array, and using the equation

$$\nu_{i,j} = \nu_{i-1,j} - \nu_{i,j+1}$$

to calculate the others. Though we present this only in the context of designs, it can be applied in any situation in which the hypotheses of (1.57) hold.

The method is particularly useful for Steiner systems: we may take $l = k$ and $\{x_1, \ldots, x_k\}$ a block, so that

$$\mu_i = \begin{cases} \lambda_i & \text{for } 0 \le i \le t, \\ 1 & \text{for } t \le i \le k. \end{cases}$$

We illustrate for the 5-(24, 8, 1) design (Table 1.1.)

759								
506	253							
330	176	77						
210	120	56	21					
130	80	40	16	5				
78	52	28	12	4	1			
46	32	20	8	4	0	1		
30	16	16	4	4	0	0	1	
30	0	16	0	4	0	0	0	1

Table 1.1.

The triangle shows immediately that, in any design with these parameters, the intersection of any two blocks has even cardinality. In particular, any 4-(23, 7, 1) design is tight, in accordance with (1.42).

Gross (1974) used the method to show the following result.

(1.55) Theorem. *Let $\mathcal{D}$ be a t-$(v,k,1)$ design. Suppose that, for some s with $0 \le s < t$, no two blocks meet in s points. Then one of the following holds:*
(a) $s = t-2$, $\mathcal{D}$ is a projective plane or an extension of one;
(b) $s = 0$, $k = t+1$, $v = 2t+3$;
(c) $s = 1$, $k = t+1$, $v = 2t+2$;
(d) $s = 0$, $t = 4$, $k = 7$, $v = 23$;
(e) $s = 1$, $t = 5$, $k = 8$, $v = 24$. □

The only known examples of designs under cases (b) and (c) are the 2-(7, 3, 1), 3-(8, 4, 1), 4-(11, 5, 1) and 5-(12, 6, 1) designs.

We now come to the final topic in this chapter, the construction of the 5-(12, 6, 1) and 5-(24, 8, 1) designs which have come up so often.

These designs are intimately related to their automorphism groups, the 5-fold transitive groups M_{12} and M_{24} discovered by Mathieu (1861), (1873). Witt (1938a,b) gave new constructions of the Mathieu groups, as transitive extensions of permutation groups of smaller degree (adding one point at a time), and used the groups to construct the designs. The procedure was reversed by Lüneburg (1969), who constructed the designs (as three-times extensions of AG(2,3) and PG(2,4) respectively), and defined the Mathieu groups as their automorphism groups. Below, we give Lüneburg's construction. The easiest way to construct (and prove uniqueness of) these designs is via coding theory, using the ternary and binary *Golay codes* associated with them. This is described in Chapter 11. Many other constructions have been given. We mention some of these below, while another is explained in Chapter 6.

Lüneburg's construction of the 5-(24, 8, 1) design is based on the following combinatorial properties of the unique projective plane of order 4, viz. PG(2,4).

(1) PG(2,4) contains exactly 168 (Type II) ovals (i.e. 6-sets meeting every line in 0 or 2 points). These fall into three classes $\mathcal{O}_1, \mathcal{O}_2, \mathcal{O}_3$ of size 56, with the property that two ovals belong to the same class if and only if their intersection has even size.

(2) PG(2,4) contains exactly 360 Baer subplanes; these are 7-sets meeting each line in 1 or 3 points (see p. 10). They fall into three classes $\mathcal{S}_1, \mathcal{S}_2, \mathcal{S}_3$ of size 120, with the property that two Baer subplanes belong to the same class if and only if their intersection has odd size.

(3) The numbering in (1) and (2) can be chosen so that, for $O \in \mathcal{O}_i$, $S \in \mathcal{S}_j$, $|O \cap S|$ is even if and only if $i = j$.

Now let $(X, \mathcal{L}) = \mathrm{PG}(2,4)$, and let $\infty_1, \infty_2, \infty_3$ be three new points. Construct a structure with point set $X \cup \{\infty_1, \infty_2, \infty_3\}$, with blocks of four types:
(a) $L \cup \{\infty_1, \infty_2, \infty_3\}$ for each $L \in \mathcal{L}$;
(b) $O \cup \{\infty_1, \infty_2, \infty_3\} \setminus \{\infty_i\}$ for each $O \in \mathcal{O}_i$, $i = 1, 2, 3$;

(c) $S \cup \{\infty_i\}$ for each $S \in \mathcal{S}_i$, $i = 1, 2, 3$;
(d) $L \triangle L'$ for $L, L' \in \mathcal{L}$, $L \neq L'$.

(Here $L \triangle L'$ is the *symmetric difference* of L and L'.)

Each block has cardinality 8, and there are

$$21 + 168 + 360 + 210 = 759 = \binom{24}{5} \Big/ \binom{8}{5}$$

blocks, so the average number of blocks containing a 5-set is 1. Some checking verifies that two blocks have at most 4 common points; so a 5-set lies in at most one block, and hence in exactly one. Thus, we have constructed a 5-(24, 8, 1) design.

The uniqueness of the design can also be shown by this method. We already noted that two blocks of such a design $\mathcal{D}$ meet in 0, 2 or 4 points. Let $\infty_1, \infty_2, \infty_3$ be three points of $\mathcal{D}$. Then the derived design $\mathcal{D}_{\infty_1\infty_2\infty_3}$ is a projective plane of order 4, necessarily isomorphic to PG(2, 4). (See Chapter 6.) Using the above observation it can be checked that blocks containing 3, 2, 1 or none of $\infty_1, \infty_2, \infty_3$ correspond to lines, ovals, Baer subplanes, and symmetric differences of pairs of lines of PG(2, 4), and that the partitions of ovals and Baer subplanes according to which points ∞_i are in the corresponding blocks is the one given.

The partitions defined in (1) and (2) above have nice descriptions in both group-theoretic and coding-theoretic terms. From a group-theoretic viewpoint, consider the group PGL(3, 4) of automorphisms of PG(2, 4) induced by non-singular linear transformations of the underlying vector space. This group acts transitively on the sets of ovals and Baer subplanes. It has a subgroup PSL(3, 4) of index 3, consisting of automorphisms induced by linear transformations of determinant 1. Both ovals and Baer subplanes fall into three orbits for this group, yielding the desired partitions. The coding-theoretic description will be given in Chapter 11.

It is possible to construct the 5-(12, 6, 1) design similarly, by extending the affine plane AG(2,3) three times, identifying appropriate geometric objects in the plane. We have outlined this method in Exercise 13. An easier approach is to find this design inside the larger one.

A *unital* in PG(2, 4) is a set of 9 points meeting any line in 1 or 3 points. It can be shown that PG(2, 4) contains 280 unitals. Let U be one of these, and set $Y = U \cup \{\infty_1, \infty_2, \infty_3\}$. Then Y meets any block of the 5-(24, 8, 1) design in an even number (at most 6) of points. So

$$(Y, \{B \cap U : B \text{ a block}, |B \cap U| = 6\})$$

is a 5-(12, 6, 1) design.

A quick construction of this design runs as follows. There is (up to equivalence) a unique Hadamard matrix of order 12. (This follows from our remarks (1.28) about the relation between Hadamard matrices and Hadamard designs, and the uniqueness of the 2-(11, 5, 2) design — see Exercise 8.) Let H denote this matrix. Any pair of rows of H agree in 6 positions and disagree in 6, the resulting pair of 6-sets being unaffected by row and column sign changes. If $\mathcal{B}$ is the collection of $2\binom{12}{2} = 132$ 6-sets which arise in this way, then $(\{1,\ldots,12\}, \mathcal{B})$ is a 5-(12, 6, 1) design.

The 2-(11, 5, 2) design can also be used to construct the 3-(22, 6, 1) design: see Exercise 8.

Many other papers describe the 5-(24, 8, 1) design. The 'Miracle Octad Generator', or MOG (Conway (1976), Curtis (1976)) is a convenient computational device for finding the block containing any five points. Todd (1966) gives a list of all 759 blocks of the design. See also Jónsson (1972), and the ATLAS *of Finite Groups* (Conway *et al.* (1985)). In Chapter 11, we will see (implicitly) many more constructions of the design (via its connection with the binary Golay code), as well as a short proof of its uniqueness.

Appendix: The Principle of Inclusion and Exclusion

Let $A_1, \ldots, A_n$ be subsets of a finite set X. The Principle of Inclusion and Exclusion asserts that the number of elements of X which lie in none of the sets A_i can be computed if we know, for each set $J \subseteq \{1,\ldots,n\}$, how many elements lie in all the sets A_j for $j \in J$ (and perhaps others). For $J \subseteq \{1,\ldots,n\}$, set

$$A(J) = \bigcap_{j \in J} A_j,$$

with the convention that $A(\emptyset) = X$.

(1.56) Theorem. *The number of points of X lying in none of $A_1, \ldots, A_n$ is*

$$\sum_{J \subseteq \{1,\ldots,n\}} (-1)^{|J|} \cdot |A(J)|.$$

PROOF. An element of X lying in none of the sets A_i is counted once in the term with $J = \emptyset$ in the sum. If x is any other element, and $K = \{j : x \in A_j\}$, $|K| = k > 0$, then the contribution of x to the sum is

$$\sum_{J \subseteq K} (-1)^{|J|} = \sum_{j=0}^{k} (-1)^j \binom{k}{j} = (1-1)^k = 0,$$

by the Binomial Theorem. □

(1.57) Corollary. *With the above notation, suppose that $|A(J)| = \lambda_j$ whenever $|J| = j$, for $j = 0, \ldots, n$. Then the number of elements lying in no set A_i is*

$$\sum_{j=0}^{n}(-1)^j\binom{n}{j}\lambda_j.$$

In fact, with the hypotheses of (1.56), the number of points lying in a prescribed collection of sets A_j and no others can be computed (see Exercise 17). This fact was used several times in the text.

Exercises

1. Let V be a vector space of dimension $n+1$ over $\mathsf{F} = \mathsf{F}_q$, where q is a prime power and $n \geq 2$. Choose an integer l with $1 \leq l < n$, and let X and $\mathcal{B}$ be the sets of 1- and $(l+1)$-dimensional subspaces of V, with incidence defined by inclusion. (These are the points and l-flats of the projective space $\mathrm{PG}(n,q)$.) Prove that $(X, \mathcal{B})$ is a 2-design, and calculate its parameters.

2. The *Kronecker product* or *tensor product* of matrices $A = (a_{ij})$ and $B = (b_{kl})$ is the matrix $A \otimes B$ with $((i,k),(j,l))$ entry $a_{ij}b_{kl}$. Prove that the Kronecker product of Hadamard matrices of orders n_1 and n_2 is a Hadamard matrix of order n_1n_2. Prove also that the Kronecker product of a number of copies of the matrix $\left(\begin{smallmatrix}1 & 1\\ 1 & -1\end{smallmatrix}\right)$ is a Sylvester matrix.

3. Suppose that H is a Hadamard matrix of order n with constant row and column sums d. Prove that $n = d^2$, with d even. Prove also that, if the -1s in H are replaced by 0s, the resulting matrix is the incidence matrix of a square 2-$(4u^2, 2u^2+u, u^2+u)$ design, where $d = 2u$. (Note: d may be positive or negative.)

4. Let $\mathcal{D}$ be a Hadamard 2-design. Prove that any line has two or three points.

Suppose now that every line has three points. Show that the dual of $\mathcal{D}$ satisfies the following condition:

(*) If B_1, B_2 are blocks and p is a point not in $B_1 \cap B_2$, then a unique block contains $B_1 \cap B_2$ and p.

Let $\mathcal{D}^\top = (X, \mathcal{B})$, where $X = \{x_1, \ldots, x_v\}$. For each $B \in \mathcal{B}$, let $\mathbf{v}(B)$ be the vector in F^v (where $\mathsf{F} = \mathsf{F}_2$), with

$$\mathbf{v}(B)_i = \begin{cases} 0 & \text{if } x_i \in B, \\ 1 & \text{otherwise.} \end{cases}$$

Let $W = \{\mathbf{v}(B) : B \in \mathcal{B}\} \cup \{\mathbf{0}\}$. Prove that W is a vector subspace of F^v.

Deduce that $\mathcal{D}$ is a projective geometry over F.

5. Let $\mathcal{D} = (X, \mathcal{B})$ be a t-(v,k,λ) design, and let $x_1, \ldots, x_{t+1}$ be points; let μ be the number of blocks containing $x_1, \ldots, x_{t+1}$. Use the Principle of Inclusion and

Exclusion to show that the number of blocks which contain none of $x_1, \ldots, x_{t+1}$ is $F + (-1)^{t+1}\mu$, where the number F depends only on the parameters of the design.

Deduce that, if $v = 2k + 1$ and t is even, then

$$\mathcal{E} = (X \cup \{\infty\}, \{B \cup \{\infty\}, X \setminus B : B \in \mathcal{B}\})$$

is an extension of $\mathcal{D}$.

6. Prove that the only square 2-design which can be extended more than once is the projective plane of order 4.

7. Use Dembowski's Theorem (1.34)(b) to prove that an extendable inversive plane has order 2, 3, 4, 8 or 13.

REMARK. An embedding theorem of Kantor (1974) shows that no inversive plane of even order greater than 2 is extendable.

8. (a) Prove that there is a unique 2-(11, 5, 2) design, up to isomorphism.

(b) Let X be the set of points and blocks of the 2-(11, 5, 2) design. Let $\mathcal{B}$ consist of 6-sets of the following three types:

- a point and the five blocks containing it;
- a block and the five points contained in it;
- an oval and its tangents.

Prove that $(X, \mathcal{B})$ is a 3-(22, 6, 1) design.

(This construction is due to Assmus, Mezzaroba and Salwach (1977).)

9. Let q be a power of 2, and let S be the conic in $\mathrm{PG}(2, q)$ with equation $xz = y^2$. Prove that the nucleus of S is $p = [(0, 1, 0)]$ (this means, the 1-dimensional subspace spanned by the vector $(0, 1, 0)$). Prove also that, if s is any point of S, then the Type I oval $S \cup \{p\} \setminus \{s\}$ is a conic if and only if $q = 2$ or $q = 4$.

(Hint: Any five points with no three collinear determine a unique conic.)

10. Using the method of intersection triangles to prove that, if Y is a set of 5 points of a 3-(10, 4, 1) design containing no block, then the complement of Y contains no block.

11. Let B be a block of a t-design $(X, \mathcal{B})$, and suppose that the cardinality of the intersection of B with the other blocks takes just s distinct values $x_1, \ldots, x_s$, where $s \le t$. Prove that, for $1 \le i \le s$,

$$(X \setminus B, \{B' \setminus B : B' \in \mathcal{B}, |B \cap B'| = x_i\})$$

is a $(t + 1 - s)$-design.

What designs are obtained from the 4-(23, 7, 1) design by this procedure?

12. Let H be a Hadamard matrix of order $4t$. Let $\mathcal{D}$ be the structure with point set $\{1, \ldots, 4t\}$, in which, for each pair of rows of $\mathcal{D}$, the set of coordinates where

these rows agree and the set where they disagree are both blocks, and these are all the blocks.

(a) Show that $\mathcal{D}$ is a 3-structure, and find its parameters.

(b) Show that, if $\mathcal{D}$ is a 4-structure, then $t = 3$.

(c) A block of $\mathcal{D}$ occurs with multiplicity at most $2t$. Prove that, if every block has multiplicity $2t$, then H is a Sylvester matrix.

13. Show that AG(2,3) contains exactly 54 (Type I) ovals. Show further that there is a unique way to partition these ovals into three classes $\mathcal{O}_1$, $\mathcal{O}_2$, $\mathcal{O}_3$ of 18 ovals each with the property that any three non-collinear points lie in a unique oval of each class.

Let X and $\mathcal{L}$ be the point and line sets of AG(2,3), and let $\infty_1, \infty_2, \infty_3$ be three new points. Define a structure $\mathcal{D}$ with point set $X \cup \{\infty_1, \infty_2, \infty_3\}$, having blocks of four types:

$L \cup \{\infty_1, \infty_2, \infty_3\}$ and $X \setminus L$ for all $L \in \mathcal{L}$;

$O \cup \{\infty_1, \infty_2, \infty_3\} \setminus \{\infty_i\}$ and $(X \setminus O) \cup \{\infty_i\}$ for all $O \in \mathcal{O}_i$, $i = 1, 2, 3$.

Prove that $\mathcal{D}$ is a 5-(12, 6, 1) design.

Show further that any 5-(12, 6, 1) design is isomorphic to $\mathcal{D}$.

14. Prove that the only extendable affine designs are affine planes.

15. (a) Show that, if π is any permutation of the point set of PG$(d, 2)$, $d > 2$, then there is a hyperplane H such that $\pi(H)$ intersects every hyperplane. [Hint: Let L be a line; then $|\pi^{-1}(L)| = 3 \leq d$, so there is a hyperplane H containing $\pi^{-1}(L)$.

(b) In the Paley design $P(q)$, where $q \equiv 3 \pmod 4$, for any block B, $-B$ is disjoint from some block.

(c) Deduce that PG$(d, 2)$ and $P(2^{d+1} - 1)$ are isomorphic only for $d = 2$.

[This argument is due to A. Blokhuis.]

16. Prove that, if a projective plane Π of order n has a subplane π of order m, then either $n = m^2$ or $n \geq m(m+1)$. Show further that $n = m^2$ if and only if every line of Π contains a point of π; and that $n = m(m+1)$ if and only if there is a line L of Π containing no point of π, such that every point of L lies on a line of π.

17. Assume the notation of the Principle of Inclusion and Exclusion (1.56). For $J \subseteq \{1, \ldots, n\}$, let $B(J)$ be the set of elements lying in the sets A_j for $j \in J$ and no others. Then (1.56) gives a formula for $|B(\emptyset)|$. Find a formula for $|B(J)|$ for arbitrary J.

18. (a) How many functions f from $\{1, \ldots, n\}$ to $\{1, \ldots, n\}$ have the property that no member of a subset J is contained in the range of f? Deduce that

$$n! = \sum_{j=0}^{n} (-1)^j \binom{n}{j} (n-j)^n.$$

(b) How many permutations of $\{1,\ldots,n\}$ fix every point in the subset J? Deduce that the number of *derangements* of $\{1,\ldots,n\}$ (permutations without fixed points) is

$$n!\sum_{j=0}^{n}\frac{(-1)^j}{j!}.$$

Show that this is the closest integer to $n!/e$.

2. Strongly regular graphs

The theory of designs concerns itself with questions about subsets of a set (or relations between two sets) possessing a high degree of regularity (although this is no guarantee of high global symmetry). By contrast, the large and amorphous area called 'graph theory' is mainly concerned with questions about 'general' relations on a set. This generality usually means that either the questions asked are too particular, or the results obtained are not powerful enough, to have useful consequences for design theory. There are some places where the two theories have interacted fruitfully; in the next six chapters, several of these areas will be considered. The unifying theme is provided by a class of graphs, the 'strongly regular graphs', introduced by Bose (1963), whose definition reflects the symmetry inherent in t-designs. First, however, we give some general definitions from graph theory, and an elementary classification theorem which will recur like a motif in subsequent chapters.

(2.1) DEFINITION. A *graph* consists of a finite set of *vertices* together with a set of *edges*, where an edge is a subset of the vertex set of cardinality 2. In the language of graph theory, our graphs are *undirected* (we do not allow edges to be ordered pairs), and without *loops* (we do not permit the two vertices comprising an edge to be equal) or *multiple edges* (a given pair of vertices can comprise at most one edge).

As with designs, there is an alternative definition: a graph consists of a finite set of vertices and a set of edges, with an 'incidence' relation between vertices and edges, having the properties that any edge is incident with exactly two vertices, and any two vertices are incident with at most one edge. Still another definition: a graph consists of a finite set of vertices together with a symmetric irreflexive binary relation (called *adjacency*) on the vertex set. We take the first definition, but use freely the language associated with the others; we say that two vertices are adjacent, or an edge is incident with a vertex.

A graph is called *complete* if every pair of vertices are adjacent, and *null* if it has no edges at all. The *complement* of the graph Γ is the graph $\overline{\Gamma}$ whose edge set is the complement of the edge set of Γ (relative to the set of all 2-element subsets of the vertex set). This is, of course, quite a different notion from the complement of a design! In the graph Γ, we let $\Gamma(x)$ denote the set of vertices adjacent to the vertex x. Given a subset S of the vertex set, $\Gamma|S$ denotes the graph with vertex set S whose edges are the edges of Γ which are contained in S. If the meaning is clear from the context, we sometimes call this graph simply S.

For example, let n be a natural number, and V the set of all $n \times n$ Latin squares with entries $\{1, 2, \ldots, n\}$. Form a graph Γ_n with vertex set V by declaring that two Latin squares are adjacent whenever they are orthogonal. As we saw in (1.31), the existence of a projective or affine plane of order n is equivalent to that of $n-1$ mutually orthogonal Latin squares, that is, of a complete subgraph of Γ_n on $n-1$ vertices. Graph theory provides some general results about the size of complete subgraphs of a graph; not surprisingly, these are far too weak to force the existence of a plane of given order except in trivial cases. This simple approach seems unlikely to produce any useful results for the theory of finite planes.

On the other hand, graphs, sometimes provide a simple construction for designs of various kinds.

(2.2) EXAMPLE. Let Γ be the complete graph on five vertices, and let E be its edge set. Let $\mathcal{B}$ be the set of subgraphs of the three shapes shown in Fig. 2.1. Then $(E, \mathcal{B})$ is a 3-(10, 4, 1) design.

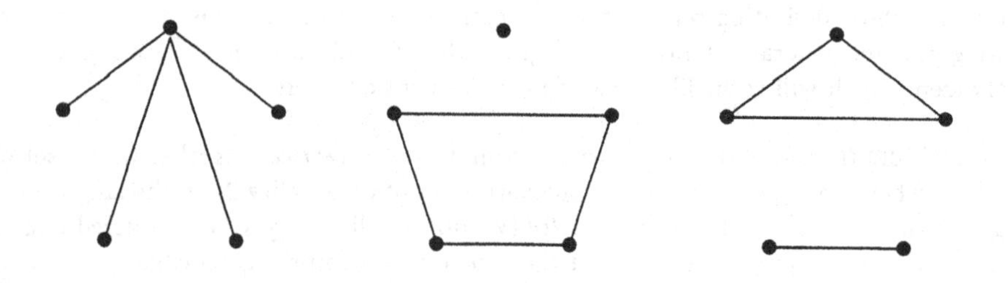

Fig. 2.1. A 3-design

If x is a vertex of a graph Γ, the *valency* of x is the number of edges containing x, or equivalently, the number $|\Gamma(x)|$ of vertices adjacent to x. If all vertices have the same valency, the graph is called *regular*, and the common valency is the *valency* of the graph. Thus, an arbitrary graph is a 0-design, with block size $k = 2$; and a regular graph is a 1-design. (The only graph which is a 2-design is the complete graph, which is sometimes referred to as the *pair design.*)

We begin with the following result, which will prove to be much more significant than it looks.

(2.3) Theorem. *Let Γ be a graph with the following property:*

(∗) *Any edge $\{x, y\}$ is contained in a triangle $\{x, y, z\}$ having the property that any further vertex is joined to exactly one of x, y and z.*

Then Γ is isomorphic to one of the following:
(a) a null graph;
(b) a 'windmill' (consisting of a number of triangles with a common vertex);
(c) three special regular graphs, having 9, 15 and 27 vertices respectively.

PROOF. Let Γ be a graph satisfying $(*)$. We may suppose that Γ has at least one edge, and that no vertex of Γ is adjacent to all others, since otherwise we are in case (a) or (b).

Let x be any vertex of Γ. Then $\Gamma|(\{x\} \cup \Gamma(x))$ is a windmill, so x has even valency, $2u$ say, with $u \geq 2$. Number the triangle containing x as $T_1, T_2, \ldots, T_u$, and let $T_i = \{x, y_{i,0}, y_{i,1}\}$.

Now let z be any vertex not equal or adjacent to x. By $(*)$, z is joined to just one of $y_{i,0}$ and $y_{i,1}$, say y_{i,ϵ_i}, for $i = 1, \ldots, u$. We can use the function $f_z : \{1, \ldots, u\} \to \{0, 1\}$ defined by

$$f_z(i) = \epsilon_i,$$

as a 'label' for the vertex z. By choice of the numbering of the neighbours of x, we can suppose that the all-zero function occurs as a label.

Let $\Delta(x) = \overline{\Gamma}(x)$ be the set of non-neighbours of x. We show next:
(1) If $z, z' \in \Delta(x)$ are adjacent, then f_z and $f_{z'}$ agree in exactly one position.
(2) If $z, z' \in \Delta(x)$ are nonadjacent but have a common neighbour in $\Delta(x)$, then f_z and $f_{z'}$ agree in all but two positions.
(3) If $z, z' \in \Delta(x)$ are non-adjacent and have no common neighbour in $\Delta(x)$, then $f_z = f_{z'}$.

Proof of (1): In this case, z and z' lie in a triangle whose third vertex is a neighbour of x, say $y_{i,\epsilon}$; so $f_z(i) = f_{z'}(i) = \epsilon$. But z and z' have no more common neighbours, so f_z and $f_{z'}$ do not agree in any other position.

Note that, for any $z \in \Delta(x)$ and any i, there is a unique neighbour z' of z for which $f_{z'}(i) = f_z(i)$.

Proof of (2): If z'' is a common neighbour, the result follows by applying (1) to the pairs $\{z, z''\}$ and $\{z'', z'\}$.

Proof of (3): If $\epsilon = f_z(i) \neq f_{z''}(i)$, then z'' is joined to a vertex of the unique triangle $\{z, y_{i,\epsilon}, z'\}$ containing z and $y_{i,\epsilon}$, necessarily z'.

Suppose that $u \geq 3$. Take $z \in \Delta(x)$ and let (z, z', z'', z''') be a path of length 3 in $\Delta(x)$ such that $f_z(i) = f_{z'}(i)$, $f_{z'}(j) = f_{z''}(j)$, $f_{z''}(k) = f_{z'''}(k)$, where i, j, k are three distinct coordinates. Then f_z and $f_{z'''}$ agree in just these three coordinates. But since (1)–(3) are exhaustive, it follows that either $u = 3$ or $u - 2 = 3$. Thus $u = 2, 3$ or 5.

In each case, we have enough information to reconstruct the graph. For example, if $u = 5$, then each function which takes the value 1 an even number of times must occur exactly once as the label of a vertex in $\Delta(x)$; two vertices of $\Delta(x)$ are adjacent precisely when their labels agree in just one coordinate. The number of vertices is $1 + 10 + 16 = 27$. This graph is called the *Schläfli graph.* (However, this name was originally used by Seidel (1968) for the complement of the graph defined here.)

For $u = 2$ and $u = 3$, there are simpler descriptions of Γ. For $u = 2$, the vertices are the $3^2 = 9$ cells of a 3×3 array, two cells adjacent whenever they belong to the same row or to the same column. For $u = 3$, the vertices are the $\binom{6}{2} = 15$ 2-element subsets of a 6-element set, two vertices adjacent when they are disjoint. □

Let us look at graphs satisfying (*) another way. Such a graph has the property that non-adjacent vertices have the same valency. (For if x lies in u triangles, that is, has valency $2u$, and z is non-adjacent to x, then z is joined to u neighbours of x, all in distinct triangles, and so z has valency at least $2u$; the same argument with x and z interchanged shows that the valencies are equal. Note incidentally that x and z have u common neighbours.) Now suppose that neither (a) nor (b) of (2.3) holds. Let x and y be vertices whose valencies are unequal. Then x and y are adjacent; let z be the third vertex of the triangle containing them. Then the valency of z differs from that of at least one of x and y, say x. Now any further vertex is adjacent to either x or y, and to either x or z, but not to both y and z; so every further vertex is adjacent to x, and the graph is a windmill, contrary to assumption.

We conclude that Γ is regular, with valency $2u$, say; two adjacent vertices have a unique common neighbour, while two non-adjacent vertices have u common neighbours.

From these conditions, we abstract the definition of a strongly regular graph.

(2.4) DEFINITION. A *strongly regular graph* with *parameters* (n, k, λ, μ) is a graph Γ with n vertices, not complete or null, in which the number of common neighbours of x and y is k, λ or μ according as x and y are equal, adjacent or non-adjacent respectively. (The condition involving equality of x and y just says that the graph is regular with valency k.)

The parameters (n, k, λ, μ) are in common use. However, this book is about the connections among graphs, codes and designs; in some of the situations we will meet, k, λ and μ have the same value in the design and in the graph, but in others they do not. (This problem is exacerbated by the fact that the parameter n is often called v.) In earlier editions of this book, we used (n, a, c, d) for the parameters of a strongly regular graph, and where necessary to avoid confusion, we adopt the same device here.

Thus, graphs which satisfy conclusion (c) of (2.3) are strongly regular, with parameters $(6u-3, 2u, 1, u)$, for $u = 2, 3, 5$ respectively. We can re-phrase (2.3) as follows:

(2.5) Proposition. *Let Γ be a strongly regular graph with parameters $(6u-3, 2u, 1, u)$. Then $u = 2, 3$ or 5, and there is a unique graph for each value of u.* □

The parameters of a strongly regular graph are not independent:

(2.6) Proposition. *If a strongly regular graph has parameters (n, k, λ, μ), then*

$$k(k-\lambda-1) = (n-k-1)\mu.$$

PROOF. Count in two ways the edges $\{y, z\}$, where $y \in \Gamma(x)$, $z \notin \Gamma(x)$. □

(2.7) Proposition. *The complement of a strongly regular graph is strongly regular.*

PROOF. Let Γ be strongly regular, with parameters (n, k, λ, μ). Clearly $\overline{\Gamma}$ is regular, with valency $\overline{k} = n-k-1$. If x and y are adjacent in $\overline{\Gamma}$, then they are non-adjacent in Γ; applying Inclusion-Exclusion (1.57) to their neighbours, we find that

$$\overline{\lambda} = (n-2) - 2k + \mu = (n-2k+\mu-2)$$

vertices are adjacent to neither in Γ. Similarly, $\overline{\mu} = n - 2k + \lambda$. □

REMARK. We obtain our first necessary conditions on the parameters of a strongly regular graph, namely $n \geq 2k - \mu + 2$ and $n \geq 2k - \lambda$.

We turn now to examples of strongly regular graphs. Only a few graphs are small enough to be drawn. In drawings such as Fig. 2.2, a small circle represents a vertex, and an arc an edge, but two arcs may 'cross' without requiring a vertex at the crossing-point. Five strongly regular graphs are displayed in Fig. 2.2. Larger graphs must be described in words.

(2.8) EXAMPLE. The *triangular graph* $T(m)$ $(m \geq 4)$ has as vertices the 2-element subsets of a set of cardinality m; two distinct vertices are adjacent if and only if they are not disjoint. $T(m)$ is strongly regular, with parameters

$$n = \tfrac{1}{2}m(m-1),\ k = 2(m-2),\ \lambda = m-2,\ \mu = 4.$$

The fifth graph in Fig. 2.2, which is called the *Petersen graph*, is the complement of $T(5)$. The *square lattice graph* $L_2(m)$ has vertex set $S \times S$, where S is a set

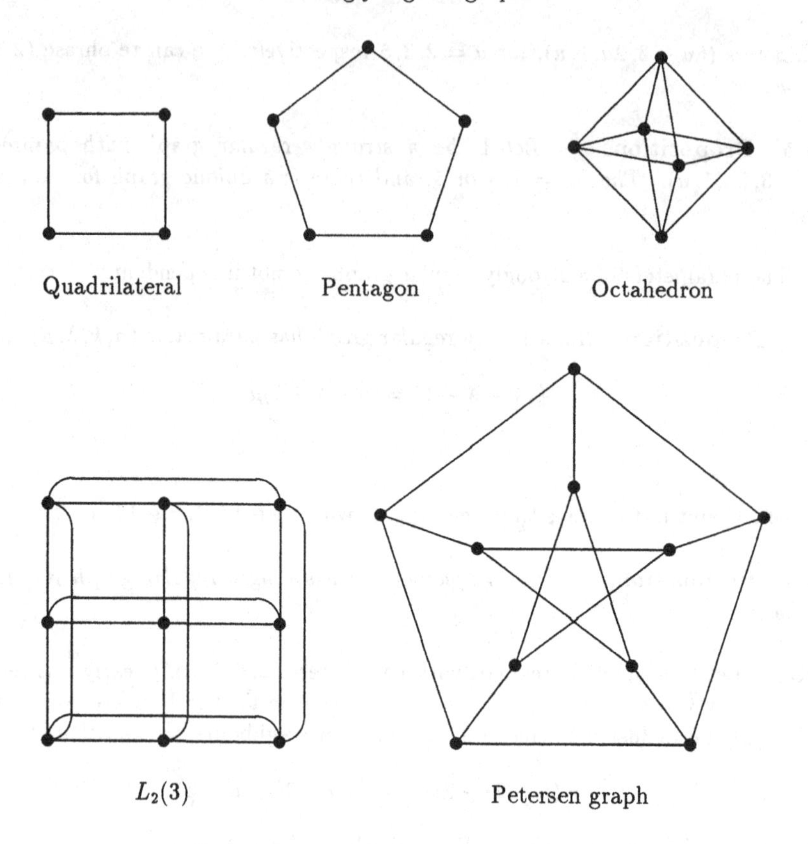

Fig. 2.2. Some strongly regular graphs

of cardinality m; two distinct vertices are adjacent if and only if they agree in one coordinate. $L_2(m)$ is strongly regular, with parameters

$$n = m^2,\ k = 2(m-1),\ \lambda = m-2,\ \mu = 2.$$

The first and fourth graphs in Fig. 2.2 are $L_2(2)$ and $L_2(3)$. However, larger triangular and square lattice graphs are more conveniently drawn as in Fig. 2.3, where the convention is that any two vertices which lie on a line are adjacent.

The first two graphs in the conclusion of (2.3) are $L_2(3)$ and the complement of $T(6)$.

(2.9) EXAMPLE. The disjoint union of r complete graphs each on m vertices ($r, m > 1$) is strongly regular, with parameters

$$n = rm,\ k = m-1,\ \lambda = m-2,\ \mu = 0.$$

This graph is denoted by $r.K_m$. Any strongly regular graph with $\mu = 0$ is of this form. The graph $r.K_2$ is called a *ladder graph*. The complement of $r.K_m$ is called a

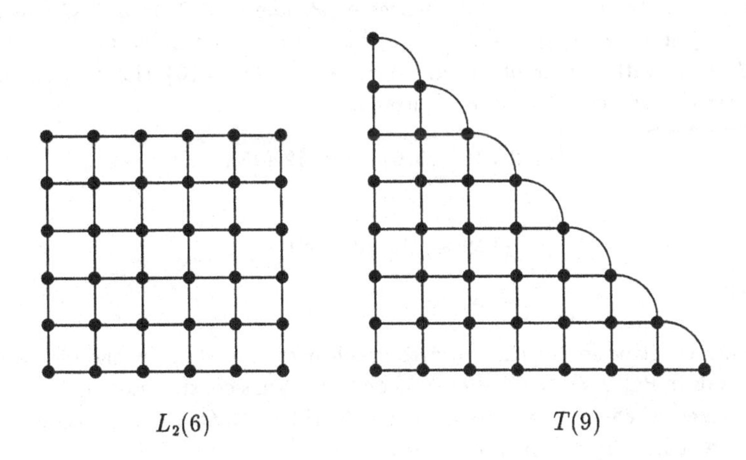

Fig. 2.3. Square lattice and triangular graphs

complete multipartite graph with block size m. The complement of a ladder graph is called a *cocktail party graph* CP(r); it models a cocktail party made up of r couples, at which each participant speaks to everybody except her/his partner. The first and third graphs in Fig. 2.2 are cocktail party graphs.

(2.10) EXAMPLE. Let q be a prime power with $q \equiv 1 \bmod 4$. The *Paley graph* $P(q)$ has as vertex set the finite field GF(q), with two vertices adjacent if and only if their difference is a non-zero square. (Since -1 is a square in GF(q), the joining rule is symmetric.) It is strongly regular, with parameters

$$n = q,\ k = \frac{1}{2}(q-1),\ \lambda = \frac{1}{4}(q-5),\ \mu = \frac{1}{4}(q-1),$$

and is isomorphic to its complement. The second and fourth graphs of Fig. 2.2 are $P(5)$ and $P(9)$. The adjacency matrix of the Paley graph is obtained from the Paley matrix defined in Chapter 1, by replacing the -1s by 0s.

(2.11) EXAMPLE. The *Clebsch graph* has as vertices all subsets of $\{1,2,3,4,5\}$ of even cardinality; two vertices are adjacent whenever (as subsets) their symmetric difference has cardinality 4. It is strongly regular with parameters $(16,5,0,2)$. The subgraph on the set of non-neighbours of a vertex is isomorphic to the Petersen graph. Similarly, the subgraph of the Schläfli graph on the set of non-neighbours of a vertex is the Clebsch graph. The Clebsch and Schläfli graphs are related to configurations in classical algebraic geometry. As with the Schläfli graph, our definition of the Clebsch graph gives the complement of Seidel's original definition.

(2.12) EXAMPLE. The *Gewirtz graph* is a strongly regular graph with parameters $(56,10,0,2)$. The following construction is due to Sims. The vertex set is $\{\infty\} \cup \mathcal{P} \cup \mathcal{Q}$, where $\mathcal{P}$ is the set of Sylow 3-subgroups of the alternating group A_6, and $\mathcal{Q}$ the set of involutions in A_6. Join ∞ to all vertices in $\mathcal{P}$; join $P \in \mathcal{P}$ to $q \in \mathcal{Q}$ whenever $q^{-1}Pq = P$; join $q_1, q_2 \in \mathcal{Q}$ whenever q_1q_2 has order 4. Combinatorially, we may identify $P \in \mathcal{P}$ with a pair of disjoint 3-subsets of $\{1, \ldots, 6\}$ (its orbits). Then typical edges of the second and third types join

$$\{\{1,2,3\},\{4,5,6\}\} \text{ to } (12)(45),$$

and

$$(12)(34) \text{ to } (23)(56),$$

respectively.

Another construction of the Gewirtz graph uses as vertex set one of the three classes of ovals in PG(2, 4) (see the remarks on Lüneburg's construction in Chapter 1), two ovals joined whenever they are disjoint. We will see in Chapter 5 a general class of strongly regular graphs which includes this example.

Let G be a group of permutations of the set V. We assume that G is *transitive* on V, that is, any point of V can be carried to any other by an element of G. There is a natural component-wise action of G on $V \times V$, given by

$$(x, y)^\pi = (x^\pi, y^\pi)$$

for $\pi \in G$. The *rank* of G is defined to be the number of orbits of G in this action. Note that the *diagonal*

$$\{(x, x) : x \in V\}$$

is one orbit, so that G is *doubly transitive* if and only if it has rank 2. Suppose that G has rank 3, and let O, O' be the two orbits other than the diagonal. Suppose further that G has even order. Then G contains an involution τ. Some pair x, y of distinct points are interchanged by τ. Suppose that $(x, y) \in O_1$. Then every pair in O_1 is interchanged by an element of G. So we can take the set of unordered pairs $\{x, y\}$ for which $(x, y) \in O_1$ as the edge set of a graph Γ on V. The fact that O_1 and O_2 are orbits implies that the number of common neighbours of two adjacent vertices, or of two non-adjacent vertices, is constant; and the transitivity of G shows that Γ is regular. So Γ is strongly regular. Such a graph is called a *rank* 3 *graph.* All the strongly regular graphs we have seen so far are actually rank 3 graphs.

All finite permutation groups of rank 3 have been determined, as a result of work by Foulser (1969), Bannai (1972), Kantor and Liebler (1982), Liebeck and Saxl (1986), and Liebeck (1987), making use of the classification of finite simple groups. Thus, at least implicitly, all rank 3 graphs are known. But, as we shall see, most strongly regular graphs are not rank 3 graphs!

Let Γ be a graph with vertex set $\{x_1, \ldots, x_n\}$. The *adjacency matrix* $A(\Gamma)$ of Γ is the $n \times n$ matrix with (i, j) entry 1 if x_i and x_j are adjacent, 0 otherwise. (Strictly speaking, different orderings of the vertex set yield different, though similar, matrices. We always presuppose an ordering of the vertices; and we speak of the (x, y) entry of $A(\Gamma)$, meaning the (i, j) entry, where $x = x_i$, $y = x_j$.)

Let Γ be strongly regular, with parameters (n, k, λ, μ), and $A = A(\Gamma)$. Then the (x, y) entry of A^2 is the number of vertices adjacent to x and y. This number is k, λ or μ according as x and y are equal, adjacent or nonadjacent. Thus:

$$A^2 = kI + \lambda A + \mu(J - I - A). \tag{2.13}$$

Here, as usual, J is the all-1 matrix. Also, since Γ is regular:

$$AJ = JA = kJ. \tag{2.14}$$

Conversely, a strongly regular graph can be defined as a graph (not complete or null) whose adjacency matrix satisfies (2.13) and (2.14).

Now A and J are commuting real symmetric matrices, and so they can be simultaneously diagonalized by an orthogonal matrix. The all-1 vector $\mathbf{j}$ is an eigenvector of both A and J, with eigenvalues k and n respectively. Applying (2.13) to this vector, we obtain

$$k(k - \lambda - 1) = (n - k - 1)\mu,$$

giving another proof of (2.6). Any vector orthogonal to $\mathbf{j}$ is an eigenvector of J with eigenvalue 0. So any further eigenvalue ρ of A satisfies:

$$\rho^2 = (k - \mu) + (\lambda - \mu)\rho. \tag{2.15}$$

The roots of this quadratic equation are denoted by r and s, with the convention that $r > s$. Thus,

$$r, s = \tfrac{1}{2}\left(\lambda - \mu \pm \sqrt{(\lambda - \mu)^2 + 4(k - \mu)}\right).$$

If r and s have multiplicities f and g respectively, then we have

$$\begin{aligned} n &= f + g + 1, \\ 0 &= \mathrm{Trace}(A) = k + fr + gs. \end{aligned}$$

These equations determine f and g, since $k = \lambda = \mu$ is not possible. Indeed, we find without difficulty:

(2.16) Theorem. *The numbers*

$$f, g = \tfrac{1}{2}\left(n - 1 \pm \frac{(n - 1)(\mu - \lambda) - 2k}{\sqrt{(\mu - \lambda)^2 + 4(k - \mu)}}\right)$$

are non-negative integers. □

This result is referred to as the *integrality condition* or *rationality condition*, and provides a powerful non-existence criterion for strongly regular graphs.

(2.17) EXAMPLE. Let us reconsider the strongly regular graphs from (2.5), with $n = 6u-3, k = 2u, \lambda = 1, \mu = u$. We have

$$f, g = \tfrac{1}{2}\left(6u - 4 \mp \frac{(6u-4)(u-1)-4u}{\sqrt{(u-1)^2+4u}}\right)$$
$$= 3u - 2 \mp \frac{(3u-1)(u-2)}{u+1}.$$

Since these quantities are integers, $u+1$ divides $(3u-1)(u-2)$. Since

$$(3u-1)(u-2) = (u+1)(3u-10) + 12,$$

$u+1$ divides 12, so that $u = 2, 3, 5$ or 11. (We exclude $u = 1$ since it corresponds to a triangle, which is not strongly regular according to our definition.) We will soon give a further necessary condition from which the non-existence of such a graph with $u = 11$ follows.

Equations (2.13) and (2.14) show that the span (over the real numbers) of the three matrices I, J, A is closed under multiplication, and hence is a commutative associative algebra. If Γ is a rank 3 graph obtained from the permutation group G, then this algebra is the *centralizer algebra* of G, that is, the set of matrices which commute with all the permutation matrices in G. The multiplicities $1, f, g$ of the eigenvalues of A are the degrees of the irreducible constituents of the permutation character of G.

To analyse the integrality conditions further, we distinguish two types of parameter sets for which f and g are integers:

Type I, with $(n-1)(\mu-\lambda) = 2k$. In this case,

$$n = 1 + \frac{2k}{\mu - \lambda} > 1 + k,$$

and so $0 < \mu - \lambda < 2$. So necessarily $\mu - \lambda = 1$, and we find that $\lambda = \mu - 1$, $k = 2\mu$, $n = 4\mu + 1$. In this case, a further necessary condition, resembling the Bruck–Ryser–Chowla Theorem (1.16), was shown by Van Lint and Seidel (1966):

(2.18) Theorem. *If a Type I strongly regular graph on n vertices exists, then n is the sum of two integer squares.* □

Thus, for example, there is no Type I strongly regular graph on 21 vertices. On the other hand, the Paley graphs are of Type I, and so such graphs exist whenever n is a prime power congruent to 1 mod 4. Examples are known for other values too, e.g. 45, 225: see Mathon (1975).

Type II, when $(\mu - \lambda)^2 - 4(k - \mu)$ is the square of an integer u, u divides $(n-1)(\mu - \lambda) - 2k$, and the quotient is congruent to $n - 1 \pmod 2$. This is the 'general case'. Note that Type I graphs are also of Type II if, and only if, n is a square. The graph $P(9) \cong L_2(3)$ is an example of this.

The above analysis can be turned into an algebraic characterization of strongly regular graphs. A graph is regular if and only if $\mathbf{j}$ is an eigenvector of A; the corresponding eigenvalue is the valency. If Γ is regular, then $\mathbf{j}^\perp$ is A-invariant; and, if A is strongly regular, then $A|\mathbf{j}^\perp$ has just two eigenvalues. Conversely, if this condition holds, then $(A - rI)(A - sI)$ is a multiple of J; this easily shows that Γ is strongly regular. Hence:

(2.19) Proposition. *The regular graph Γ with adjacency matrix A is strongly regular if and only if $A|\mathbf{j}^\perp$ has just two eigenvalues.* □

We have seen that the parameters of a strongly regular graph determine the eigenvalues of A and their multiplicities. What about the converse? The eigenvalues do determine all the parameters. For r and s are the roots of the quadratic $\rho^2 = (k - \mu) + (\lambda - \mu)\rho$; so we have

$$\begin{aligned} r + s &= \lambda - \mu, \\ rs &= -(k - \mu), \end{aligned}$$

whence $\lambda = k + r + s + rs$, $\mu = k + rs$. It follows that the eigenvalues determine the multiplicities. Sometimes, it proves convenient to use these expressions for the parameters in terms of k, r, s.

In general, knowledge of the multiplicities alone does not determine the parameters. Sometimes it may give strong partial information, as the following result, due to Wielandt (1964), Chapter 5, shows. (Wielandt's argument is given for rank 3 graphs, but extends to arbitrary strongly regular graphs.)

(2.20) Theorem. *Suppose that Γ is a strongly regular graph on $n = 2m$ vertices, whose eigenvalues have multiplicities $1, m-1, m$. Then either*
(a) Γ or its complement is a ladder graph; or
(b) Γ or its complement has parameters $n = 4s^2 + 4s + 2$, $k = s(2s+1)$, $\lambda = s^2 - 1$, $\mu = s^2$, for some positive integer s.

PROOF. We may assume that $k < m$, by replacing the graph by its complement if necessary (since the complement has valency $2m - k - 1$). The matrices A, A^2, A^3 have diagonal entries $0, k, k\lambda$ respectively, and so their traces are $0, nk, nk\lambda$ respectively. Thus,

$$\begin{aligned} k + (m-1)r + ms &= 0, \\ k^2 + (m-1)r^2 + ms^2 &= 2mk, \\ k^3 + (m-1)r^3 + ms^3 &= 2mk\lambda. \end{aligned}$$

The *Perron–Frobenius Theorem* (see Gantmacher (1959)) asserts that, if A is a non-negative matrix, then A has an eigenvector with positive components, and if k is the eigenvalue associated with this eigenvector, then $|\rho| \leq k$ for all eigenvalues ρ. In our case, this gives $|r| \leq k$. (A weaker result, sufficient for our needs, is in Exercise 14.)

The first equation above shows that $k \equiv r \pmod{m}$. So either $r = k$, or $r = k - m$. In the first case, $s = -k$, $2mk^2 = 2mk$, so $k = 1$ and (a) holds. In the second case, we find $k = m - 1 - s$, $r = -1 - s$; the second equation then yields $m = 2s^2 + 2s + 1$, and the value of λ follows from the third equation. □

The only graph having the parameters of (2.20)(b) with $s = 1$ is the Petersen graph. However, there is a family of examples, due to Delsarte and Goethals (1975) and to Turyn (1974), showing that examples exist whenever $2s + 1$ is a prime power. The examples are related to the so-called *symmetric conference matrices* studied by Delsarte, Goethals and Seidel (1971).

(2.21) EXAMPLE. Let V be a 2-dimensional vector space over $\mathbf{F}_q$, where q is an odd prime power. There are $q + 1$ 1-dimensional subspaces of V; partition these into two classes P and N, each of cardinality $\frac{1}{2}(q+1)$, in any manner. Form the graph with vertex set V, in which x and y are joined whenever $[x - y] \in P$. This graph is strongly regular, with $n = q^2$, $k = \frac{1}{2}(q^2 - 1)$, $\lambda = \frac{1}{4}(q^2 - 5)$, $\mu = \frac{1}{4}(q^2 - 1)$. (These are the same parameters as the Paley graph $P(q^2)$; and indeed, the Paley graph is a special case.)

Next, choose a member of N, and select any $\frac{1}{2}(q-1)$ of its cosets; let X be the the set of $\frac{1}{2}q(q-1)$ vertices contained in these cosets. Delete each edge $\{x, y\}$ with $x \in X$, $y \notin X$, and add new edges $\{x, y\}$ for all previously non-joined edges of this form. Finally, add a new vertex ∞ joined to every vertex in X. The resulting graph is strongly regular, and has the parameters of (2.20)(b), with $q = 2s + 1$. □

The above 'switching' construction will be considered further in Chapter 4.

We now give some more necessary conditions on the parameters of strongly regular graphs. The first is due to Delsarte, Goethals and Seidel (1977). It uses an elementary lemma which will be very useful to us in the next chapter.

(2.22) Lemma. *Let $A = (a_{ij})$ be a positive definite symmetric real $n \times n$ matrix of rank d. Then there are vectors $\mathbf{v}_1, \ldots, \mathbf{v}_n \in \mathbf{R}^d$ such that $a_{ij} = \langle \mathbf{v}_i, \mathbf{v}_j \rangle$ for $i, j = 1, \ldots, n$.*

PROOF. By the 'reduction of a quadratic form to a sum of squares' (see, for example, Cohn (1974), p. 189), there is an invertible real matrix P such that

$$PAP^{\top} = \begin{pmatrix} I_d & O \\ O & O \end{pmatrix}.$$

Putting $Q = P^{-1}$, we have

$$A = Q\begin{pmatrix} I_d & O \\ O & O \end{pmatrix} Q^{\top} = Q_1 I_d Q_1^{\top},$$

where Q_1 is the $n \times d$ matrix consisting of the first d columns of Q. Now take $\mathbf{v}_1, \ldots, \mathbf{v}_n$ to be the rows of Q_1. □

A is called the *Gram matrix* of the set $\mathbf{v}_1, \ldots, \mathbf{v}_n$ of vectors. A set of vectors is uniquely determined, up to isometry of $\mathbf{R}^d$, by its Gram matrix.

(2.23) Theorem. *Let Γ be a strongly regular graph on n vertices, having the properties that Γ and $\overline{\Gamma}$ are both connected, and that the adjacency matrix of Γ has an eigenvalue of multiplicity f ($f > 1$). Then $n \leq \frac{1}{2}f(f+3)$.*

PROOF. The adjacency matrix $A = A(\Gamma)$ has three distinct eigenspaces, and any matrix having these eigenspaces is a linear combination of I, A, and $J - I - A$. In particular, there is such a linear combination E having eigenvalue 1 on the given f-dimensional eigenspace and 0 on its complement. Then E is positive semidefinite, and so is the Gram matrix of a set S of vectors in $\mathbf{R}^f$. Since $E = \alpha I + \beta A + \gamma(J - I - A)$, any vector in S has length $\sqrt{\alpha}$, and two vectors in S make an angle $\cos^{-1}(\beta/\alpha)$ or $\cos^{-1}(\gamma/\alpha)$. The vectors are all distinct, since neither Γ nor its complement is a disjoint union of complete graphs. We may normalize to assume that $\alpha = 1$, that is, S is a subset of the unit sphere Ω.

For $\mathbf{v} \in S$, let $f_{\mathbf{v}} : \Omega \to \mathbf{R}$ be the function defined by

$$f_{\mathbf{v}}(\mathbf{x}) = \frac{(\langle \mathbf{v}, \mathbf{x}\rangle - \beta)(\langle \mathbf{v}, \mathbf{x}\rangle - \gamma)}{(1-\beta)(1-\gamma)}.$$

Now $f_{\mathbf{v}}$ is a polynomial function of degree 2; and the functions $f_{\mathbf{v}}$, for $\mathbf{v} \in S$, are linearly independent, since

$$f_{\mathbf{v}}(\mathbf{w}) = \begin{cases} 1 & \text{if } \mathbf{v} = \mathbf{w}, \\ 0 & \text{if } \mathbf{v} \neq \mathbf{w}. \end{cases}$$

But these functions live in the space spanned by the f linear and $\frac{1}{2}f(f+1)$ homogeneous quadratic functions on Ω. (Constants are not required, since

$$x_1^2 + \ldots + x_f^2 = 1$$

on Ω.) Thus,

$$n = |S| \leq f + \tfrac{1}{2}f(f+1) = \tfrac{1}{2}f(f+3),$$

as required. □

Note that the heart of this result is a bound for the number of vectors which make only two distinct angles with each other. Delsarte *et al.* call this the *absolute bound*, to contrast it with other bounds which depend explicitly on the values of the angles.

(2.24) EXAMPLE. The absolute bound excludes the possibility of a strongly regular graph with parameters $(6u-3, 2u, 1, u)$, $u = 11$, left open in Example (2.17).

The next bound is called the *Krein condition*: its first proof (for rank 3 graphs) by Scott (1977), used a result of Krein on locally finite groups. The present proof is more elementary, depending on the following fact. Let A and B be $n \times n$ matrices. Their *Hadamard product* $A \circ B$ is the matrix with (i, j) entry $a_{ij}b_{ij}$.

(2.25) Lemma. *Let A and B be positive semi-definite real symmetric matrices. Then $A \circ B$ is positive semi-definite.*

PROOF. The tensor or Kronecker product $A \otimes B$ (see Exercise 2 of Chapter 1) of positive semi-definite matrices is positive semi-definite: its eigenvalues are all products of an eigenvalue of A and an eigenvalue of B. Now $A \circ A$ is a principal submatrix of $A \otimes A$, and so represents the restriction of $A \otimes A$ to a subspace. Thus $A \circ A$ is positive semi-definite. □

Now we can state the Krein conditions:

(2.26) Theorem. *Let Γ be a strongly regular graph, such that Γ and $\overline{\Gamma}$ are connected. Let Γ have eigenvalues k, r, s. Then*
(a) $(r+1)(k+r+2rs) \leq (k+r)(s+1)^2$;
(b) $(s+1)(k+s+2rs) \leq (k+s)(r+1)^2$.

PROOF. The idempotent matrix E of the proof of (2.23) is positive semi-definite, and so $E \circ E$ is positive semi-definite. But

$$E \circ E = \alpha^2 I + \beta^2 A + \gamma^2 (J - I - A),$$

so its eigenvalues can be found. Some calculation gives the result. □

There are two classes of strongly regular graphs which bear a close relationship to square designs with special polarities (cf. (1.18)).

Let $\mathcal{D} = (X, \mathcal{B})$ be a square 2-(v, k, λ) design, with $2 \leq k \leq v-2$ (to avoid trivial cases). Let σ be a polarity of $\mathcal{D}$. A point x is *absolute* if $x \in x^\sigma$. We form a graph Γ *associated with* σ on the vertex set X by joining x to y whenever $x \neq y$ and $y \in x^\sigma$. (The definition of a polarity implies that

$$x \in y^\sigma \Leftrightarrow y \in x^\sigma,$$

so the joining rule is symmetric.) The graph Γ has the property that $\Gamma(x) = x^\sigma \setminus \{x\}$ for all x. In general, Γ is not regular, since absolute points have valency $k-1$ while non-absolute points have valency k. However, if one or other type of point is absent, then Γ is strongly regular. Note that the adjacency matrix of the graph is equal to the symmetric incidence matrix given by (1.19), with all diagonal 1s replaced by 0s.

(2.27) Proposition. *(a) The graph Γ is associated with a polarity of a square design with no absolute points if and only if it is strongly regular with $\mu = \lambda$.*
(b) Γ is associated with a polarity of a square 2-design with every point absolute if and only if it is strongly regular with $\mu = \lambda + 2$.

PROOF. If no point is absolute, then $x^\sigma \cap y^\sigma = \Gamma(x) \cap \Gamma(y)$ has cardinality λ, whether or not x and y are adjacent. Conversely, if Γ is strongly regular with $\mu = \lambda$, then $(X, \{\Gamma(x) : x \in X\})$ is a square 2-design and $\sigma : x \mapsto \Gamma(x)$ a polarity without absolute points.

The other case is proved similarly, or by applying (a) to the complementary design and graph. Note that a polarity of a projective plane must have non-absolute points. □

A strongly regular graph with parameters (v, k, λ, λ) is called a (v, k, λ)*-graph*.

(2.28) Proposition. *For fixed λ, there are only finitely many (v, k, λ) graphs.*

PROOF. Let Γ be such a graph. Γ must have Type II with respect to the integrality condition. From this, we conclude that $k = \lambda + u^2$ for some integer u which divides λ. Hence $u \leq \lambda$, whence $k \leq \lambda(\lambda + 1)$, and $v \leq \lambda^2(\lambda + 2)$. □

The extremal case $v = \lambda^2(\lambda + 2)$, $k = \lambda(\lambda + 1)$ occurs for all prime power values of λ. We will see examples in Chapter 7. Note that $L_2(4)$ is a (16, 6, 2) graph. In Chapter 4. we will see that there is just one further (16, 6, 2) graph (up to isomorphism), the Shrikhande graph. It turns out that these graphs are associated with different polarities of the same 2-(16, 6, 2) design (see Exercise 5 of Chapter 4).

In the other case of strongly regular graphs associated with polarities, viz. those with $\mu = \lambda + 2$, no such finiteness theorem is known, even for graphs with $\lambda = 0$, $\mu = 2$. Three such graphs are known:
(a) $CP(2)$, with parameters (4, 2, 0, 2);
(b) the Clebsch graph, with parameters (16, 5, 0, 2);
(c) the Gewirtz graph, with parameters (56, 10, 0, 2).
All these graphs are uniquely determined by their parameters. They are associated with polarities of the trivial 2-(4, 3, 2) design, a 2-(16, 6, 2) design (the same one we already saw), and a 2-(56, 11, 2) design. It is known that there are exactly three non-isomorphic 2-(16, 6, 2) designs, and at least four non-isomorphic 2-(56, 11, 2) designs; but the other designs with these parameters do not admit polarities with every point absolute.

We conclude this chapter by mentioning some developments of (2.3) which don't fit in elsewhere.

(2.29) EXAMPLE. Let V be a vector space over the field $\mathsf{F} = \mathsf{F}_2$, and let Q be a quadratic form on V (a function from V to F which is homogeneous of degree 2 in the coordinates). We obtain a bilinear form by *polarizing* Q:

$$f(\mathbf{x}, \mathbf{y}) = Q(\mathbf{x} + \mathbf{y}) - Q(\mathbf{x}) - Q(\mathbf{y}).$$

Now Q is called *non-singular* if the only vector $\mathbf{x}$ satisfying $Q(\mathbf{x}) = 0$ and $f(\mathbf{x}, \mathbf{y}) = 0$ for all $\mathbf{y} \in V$ is $\mathbf{x} = \mathbf{0}$.

From a non-singular quadratic form Q, we obtain a graph Γ, with vertex set

$$\{\mathbf{x} \in V \;:\; \mathbf{x} \neq \mathbf{0},\ Q(\mathbf{x}) = 0\},$$

in which $\mathbf{x}$ and $\mathbf{y}$ are adjacent if and only if $f(\mathbf{x}, \mathbf{y}) = 0$.

It can be shown that Γ is strongly regular. Furthermore, Γ has the *triangle property*, a generalization of $(*)$ of (2.3): any edge $\{\mathbf{x}, \mathbf{y}\}$ is contained in a triangle $\{\mathbf{x}, \mathbf{y}, \mathbf{z}\}$ having the property that any further vertex is joined to one or all of $\mathbf{x}, \mathbf{y}, \mathbf{z}$. (If $Q(\mathbf{x}) = Q(\mathbf{y}) = f(\mathbf{x}, \mathbf{y}) = 0$, then $Q(\mathbf{x}+\mathbf{y}) = 0$; $\mathbf{x}+\mathbf{y}$ is the required third vertex of the triangle. The triangle property is a consequence of the equation

$$f(\mathbf{x}, \mathbf{w}) + f(\mathbf{y}, \mathbf{w}) + f(\mathbf{x} + \mathbf{y}, \mathbf{w}) = 0.)$$

Shult (1972b) and Seidel (1973) proved the following converse. (Shult gave the result for regular graphs only.)

(2.30) Theorem. *A non-null graph with the triangle property, in which no vertex is joined to all others, is obtained as described above from a non-singular quadratic form over* GF(2). □

This result was further generalized by Buekenhout and Shult (1974), who replaced 'triangles' by 'complete subgraphs of arbitrary size'. But it would take us too far afield even to state their result.

Exercises

1. Let Γ be one of the graphs of (2.3)(c), with vertex set V, and let ∞ be a point not in V. Set $X = V \cup \{\infty\}$, and let

$$\begin{aligned}\mathcal{B} = \{&\{\infty\} \cup T : T \text{ a triangle in } \Gamma\} \\ &\cup \{T_1 \Delta T_2 : T_1, T_2 \text{ triangles of } \Gamma \text{ with a common vertex}\}.\end{aligned}$$

Prove that $(X, \mathcal{B})$ is a 2-design.

2. Prove that there is no strongly regular graph with parameters $(28, 9, 0, 4)$. Can you do this by 'elementary' reasoning?

3. Find the 'feasible' parameter sets of strongly regular graphs on at most 20 vertices which satisfy the integrality condition. Are any of them excluded by the absolute bound or the Krein condition? For which can you construct a graph?

4. (i) Prove that, if a strongly regular graph is isomorphic to its complement, then its parameters are of Type I.

(ii) Prove that $L_2(3)$ is isomorphic to its complement.

5. Construct examples of strongly regular graphs Γ_1, Γ_2 on the same number of vertices, whose eigenvalues have the same multiplicities, such that Γ_2 is not isomorphic to Γ_1 or its complement.

6. Verify the assertions in Example (2.21).

7. Prove that the three graphs of (2.3)(c) are obtained from the non-singular quadratic forms

$$x_1x_2 + x_3x_4,$$

$$x_1x_2 + x_3x_4 + x_5^2,$$

$$x_1x_2 + x_3x_4 + x_5^2 + x_5x_6 + x_6^2$$

over $\mathbb{F}_2$ respectively, by the construction of Example (2.29).

8. (The *Friendship Theorem.*) Let Γ be a graph in which any two vertices have a unique common neighbour.

(i) Prove that either Γ is a windmill, or Γ is regular (and hence strongly regular).

(ii) Use the integrality condition to show that there is no strongly regular graph satisfying the hypothesis.

REMARK. This result is due to Erdős, Rényi and Sós (1966).

9. The *diameter* of a connected graph Γ is the maximum distance between two vertices of Γ; the *girth* of Γ is the number of vertices in the shortest cycle in Γ.

Let Γ be a regular graph with valency $k > 1$, diameter d, girth g, and n vertices. Let

$$f(k, d) = 1 + k + k(k-1) + \ldots + k(k-1)^{d-1}.$$

Prove:

(i) $g \leq 2d + 1$;

(ii) $n \geq f(k, d)$;

(iii) $n \leq f(k, \lfloor \frac{1}{2}(g-1) \rfloor)$;

(iv) if one of these bounds is met, then all are met.

(A graph meeting the bounds (i)–(iii) is called a *Moore graph* of diameter d.)

10. Prove that, if a Moore graph of diameter 2 has valency k, then $k = 2, 3, 7$ or 57.

REMARK. The unique examples with $k = 2$ and $k = 3$ are the pentagon and the Petersen graph respectively. In Chapter 6, we shall construct and prove the uniqueness of a Moore graph of valency 7, the *Hoffman–Singleton graph*. The existence of a Moore graph of valency 57 is undecided. Bannai and Ito (1973) and Damerell (1973) showed that the only Moore graph with diameter $d > 2$ is the cycle of length $2d + 1$ (with valency 2). We shall discuss the context of this result in Chapter 17.

11. Let X be the edge set of the complete bipartite graph $K_{7,11}$, and let $\mathcal{B}$ be the set of subsets of X which (as subgraphs) are isomorphic to the graph of Fig. 2.4. Prove that $(X, \mathcal{B})$ is a 3-$(77, 20, \lambda)$ design for some λ. (This example is taken from Cameron and Praeger (to appear).)

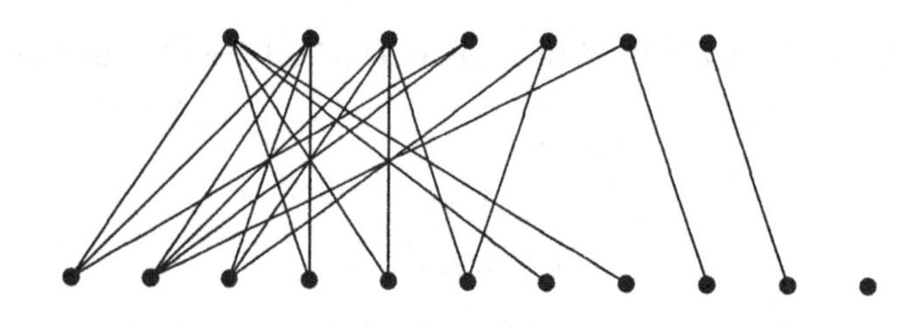

Fig. 2.4. A 3-design

12. Let $\mathcal{D}$ be a biplane (a square 2-design with $\lambda = 2$) and Σ the set of all polarities of $\mathcal{D}$ with every point absolute.

(a) Show that any $\sigma \in \Sigma$ is uniquely determined by a single pair (x, x^σ).

(b) Show that any two elements of Σ commute, and their product is a fixed-point-free automorphism of order 2.

(c) Show that the product of any three elements of Σ is a polarity with no absolute points. Deduce that, if $|\Sigma| \geq 3$, then $\mathcal{D}$ is a 2-(4, 3, 2) or 2-(16, 6, 2) design, with $|\Sigma| = 3$ or 6 respectively.

13. This alternative approach to (2.3) was suggested by J. I. Hall. Let Γ be a graph satisfying $(*)$ of (2.3). Suppose that Γ contains an edge, and that no vertex of Γ is adjacent to all others.

(1) Prove that, if x and y are vertices and $\{x\} \cup \Gamma(x) = \{y\} \cup \Gamma(y)$, then $x = y$.

(2) Let V be the vector space, over $\mathsf{F} = \mathsf{F}_2$, of functions from the vertex set X to F. For each $x \in X$, define $f_x \in V$ by

$$f_x(y) = \begin{cases} 0 & \text{if } y \in \{x\} \cup \Gamma(x) \\ 1 & \text{otherwise.} \end{cases}$$

Prove that the map $\phi : X \to V$ given by

$$\phi(x) = f_x$$

is one-to-one, that its image does not contain the 0 vector, and that

$$\{x, y, z\} \text{ is a triangle } \Leftrightarrow \phi(x) + \phi(y) + \phi(z) = 0.$$

As a consequence, we may assume that X itself is a subset of V.

(3) Deduce that the intersection of X with any subspace of V carries a subgraph satisfying (*). In particular, Γ is 'locally finite'; that is, any finite subset of X is contained in a finite subgraph which satisfies (*). So, to show that no infinite graphs can occur, it suffices to bound the size of finite graphs which can occur.

(4) Show that a quadrilateral in Γ lies in a 9-vertex subgraph isomorphic to $L_2(3)$. (If $x_1x_2x_3x_4$ is a quadrilateral, the other vertices are $x_i + x_{i+1}$ ($i = 1, \dots 4$; subscripts mod 4) and $x_1 + x_2 + x_3 + x_4$.)

(5) Let Y_1 be such a 9-vertex subgraph. Show that, for $x \notin Y_1$, $Y_1 \cup \{x\}$ lies in a 15-vertex subgraph isomorphic to the complement of $T(6)$.

(6) Let Y_2 be such a 15-vertex subgraph. Show that, for $x \notin Y_2$, $Y_2 \cup \{x\}$ lies in a 27-vertex subgraph isomorphic to the Schläfli graph.

(7) Let Y_3 be such a 27-vertex subgraph. Show that there is no vertex outside Y_3.

14. Let A be a real symmetric matrix with non-negative entries, having row and column sums k, and let ρ be an eigenvalue of A. Show that $|\rho| \leq k$. [HINT: Let $\mathbf{v}A = \rho\mathbf{v}$, where $\mathbf{v} = (v_1, \dots, v_n)$; let $|v_i| = \max\{|v_1|, \dots, |v_n|\}$, where $v_i > 0$ w.l.o.g. Show that $(\mathbf{v}A)_i \leq kv_i$.]

3. Graphs with least eigenvalue −2

In this chapter, we will give a structure theorem, due originally to Hoffman (1977), for graphs whose adjacency matrix has least eigenvalue -2 or greater. Our presentation follows Cameron, Goethals, Seidel and Shult (1976).

We begin, tangentially, by considering a problem in Euclidean geometry. A *line system* will here denote a set S of lines through the origin in some Euclidean space $\mathbf{R}^d$ having the property that any two lines in S make an angle of $90°$ or $60°$ with each other.

A *star* is a system of three lines in the plane $\mathbf{R}^2$, mutually at $60°$. A line system S is *star-closed* if, whenever two lines of S make an angle of $60°$, the third line of the star in their span also belongs to S.

(3.1) Lemma. *Any line system S is contained in a star-closed line system.*

PROOF. Choose a spanning vector of length $\sqrt{2}$ for each line in S. These vectors $\mathbf{x}$ satisfy $\langle \mathbf{x}, \mathbf{x} \rangle = 2$, $\langle \mathbf{x}, \mathbf{y} \rangle = -1, 0$ or 1 for $\mathbf{x} \neq \mathbf{y}$. Now suppose that $\langle \mathbf{x}, \mathbf{y} \rangle = -1$, so that the third line of the star is spanned by $\mathbf{x} + \mathbf{y}$. Let $[\mathbf{w}]$ be another line of S. Then $\langle \mathbf{w}, \mathbf{x} \rangle, \langle \mathbf{w}, \mathbf{y} \rangle \in \{-1, 0, 1\}$. We claim that also $\langle \mathbf{w}, \mathbf{x} + \mathbf{y} \rangle \in \{-1, 0, 1\}$. If not then, without loss, $\langle \mathbf{w}, \mathbf{x} + \mathbf{y} \rangle = 2$; but then

$$\langle \mathbf{w} - \mathbf{x} - \mathbf{y}, \mathbf{w} - \mathbf{x} - \mathbf{y} \rangle = 0,$$

so $\mathbf{w} = \mathbf{x} + \mathbf{y}$.

So $[\mathbf{x} + \mathbf{y}]$ makes an angle of $90°$ or $60°$ with all lines of S, and it may be adjoined to S if it is not already in S. After finitely many steps, we reach a star-closed line system. □

(3.2) REMARK. S is star-closed if and only if the vectors of fixed length lying on the lines of S form a *root system*, as in the theory of simple Lie algebras over $\mathbf{C}$.

A line system S is *indecomposable* if its lines are not contained in the union of two non-zero perpendicular subspaces. We now give some examples of indecomposable star-closed line systems. In what follows, $\{\mathbf{e}_i\}$ denotes a set of mutually perpendicular unit vectors.

(3.3) EXAMPLE.

$$A_n = \{[\mathbf{e}_i - \mathbf{e}_j] : 0 \le i < j \le n\}.$$

This system is contained in real n-space, because all the spanning vectors are orthogonal to the vector $\mathbf{e}_0 + \ldots + \mathbf{e}_n$.

(3.4) EXAMPLE.

$$D_n = \{[\pm\mathbf{e}_i \pm \mathbf{e}_j] : 1 \le i < j \le n\}.$$

(3.5) EXAMPLE. E_8: This system has several descriptions.

(a) $A_8 \cup \{[\mathbf{e}_i + \mathbf{e}_j + \mathbf{e}_k - \frac{1}{3}\mathbf{j}] : \{i,j,k\} \subseteq \{0,\ldots,8\}\}$,
where $\mathbf{j} = \mathbf{e}_0 + \ldots + \mathbf{e}_8$.

(b) $D_8 \cup \left\{\left[\sum_{i=0}^{8} \epsilon_i \mathbf{e}_i\right] \ : \ \epsilon_i = \pm 1,\ \prod_{i=0}^{8} \epsilon_i = 1\right\}$.

(c) $\{[\mathbf{e}_i] : 1 \le i \le 8\} \cup \{[\frac{1}{2}(\pm\mathbf{e}_i \pm \mathbf{e}_j \pm \mathbf{e}_k \pm \mathbf{e}_l)] : \{i,j,k,l\} \in \mathcal{B}\}$,
where $(\{1,\ldots,8\},\mathcal{B})$ is the unique 3-(8, 4, 1) design (the 3-dimensional affine space over $\mathbf{F}_2$).

(3.6) EXAMPLE. E_7 and E_6 are the subsets of E_8 orthogonal to a fixed line and a fixed star respectively.

(3.7) Theorem. *Any indecomposable star-closed system of lines at 90° and 60° is isomorphic to A_n $(n \ge 1)$, D_n $(n \ge 4)$, E_8, E_7 or E_6.*

PROOF. We outline the proof of this theorem. Let S be such a system. Note that any line is orthogonal to one or all members of a star, by the proof of (3.1). Let $\{[\mathbf{x}],[\mathbf{y}],[\mathbf{z}]\}$ be a star, and let $A_\mathbf{x}$, $A_\mathbf{y}$, $A_\mathbf{z}$, B be the sets of lines orthogonal to only $\mathbf{x}$, only $\mathbf{y}$, only $\mathbf{z}$, or all three. We may assume that $\mathbf{x}+\mathbf{y}+\mathbf{z} = \mathbf{0}$, and we can choose spanning vectors for the other lines so that, for example, if $[\mathbf{w}] \in A_\mathbf{x}$, then $\langle \mathbf{w},\mathbf{y}\rangle = 1$, $\langle \mathbf{w},\mathbf{z}\rangle = -1$. We always adopt this convention in what follows.

Form a graph Γ with vertex set $A_\mathbf{x}$, in which two lines are joined if and only if they are perpendicular.

Claim 1. The spanning vectors of any two lines of $A_\mathbf{x}$ have non-negative inner product.

PROOF. If $[\mathbf{u}],[\mathbf{v}] \in A_\mathbf{x}$ and $\langle \mathbf{u},\mathbf{v}\rangle = -1$, then $[\mathbf{u}+\mathbf{v}] \in S$ by star-closure; but $\langle \mathbf{u}+\mathbf{v},\mathbf{y}\rangle = 2$, a contradiction.

Claim 2. Γ satisfies the hypothesis $(*)$ of (2.3).

PROOF. Let $[\mathbf{u}]$ and $[\mathbf{v}]$ be adjacent. Then $[\mathbf{y}-\mathbf{u}],[\mathbf{z}+\mathbf{v}] \in S$ by star-closure, and $\langle \mathbf{y}-\mathbf{u},\mathbf{z}+\mathbf{v}\rangle = 1$; so $\mathbf{w} = \mathbf{y}-\mathbf{z}-\mathbf{u}-\mathbf{v} \in S$. Checking inner products shows that indeed $[\mathbf{w}] \in A_{\mathbf{x}}$. Now $\mathbf{u}+\mathbf{v}+\mathbf{w} = \mathbf{y}-\mathbf{z}$, and so $\langle\mathbf{t},\mathbf{u}\rangle + \langle\mathbf{t},\mathbf{v}\rangle + \langle\mathbf{t},\mathbf{w}\rangle = 2$ for all $[\mathbf{t}] \in A_{\mathbf{x}}$. By Claim 1, exactly one of $\langle\mathbf{t},\mathbf{u}\rangle$, $\langle\mathbf{t},\mathbf{v}\rangle$, $\langle\mathbf{t},\mathbf{w}\rangle$ is zero. So $\{[\mathbf{u}],[\mathbf{v}],[\mathbf{w}]\}$ is the required triangle.

Claim 3. Γ determines S uniquely.

PROOF. By Claim 1, the graph structure of Γ determines the Gram matrix, and so the embedding of $\{[\mathbf{x}],[\mathbf{y}],[\mathbf{z}]\} \cup A_{\mathbf{x}}$, up to isometry; and S is the 'star-closure' of this set, as in (3.1).

Now it is readily checked that, for the line systems A_n, D_n, E_6, E_7, E_8, the graph Γ is isomorphic respectively to a null graph, a windmill, or one of the three graphs of (2.3)(c). Thus, (2.3) and the above Claims complete the proof of the Theorem. □

Now let Γ be a graph whose adjacency matrix $A = A(\Gamma)$ has smallest eigenvalue -2 or greater. Then $A+2I$ is positive semi-definite and symmetric. By (2.22), it is the Gram matrix of a set $\{\mathbf{x}_1,\ldots,\mathbf{x}_n\}$ of vectors in Euclidean space. These vectors satisfy $\langle\mathbf{x}_i,\mathbf{x}_i\rangle = 2$, $\langle\mathbf{x}_i,\mathbf{x}_j\rangle = 0$ or 1 for $i \neq j$, and so they span a system of lines with angles 90° or 60°. (Now, the relevance of line systems to the topic of this chapter is clear!) The decomposition of the system into minimal pairwise orthogonal subsystems corresponds to the decomposition of the graph Γ into connected components. So, if Γ is connected (as we will assume), then the system is indecomposable, and its star-closure S is one of those given by (3.7). We say that Γ is *represented* in the system S.

So we can determine Γ by looking within the systems A_n, D_n and E_n, for sets of vectors with all inner products non-negative; adjacency in the graph corresponds to positive inner product (just the reverse of the situation in the proof of (3.7)!). In fact, since $A_n \subset D_{n+1}$ and $E_6 \subset E_7 \subset E_8$, it suffices to consider D_n and E_8; but we will tackle the system A_n first as an illustration.

Let Γ_0 be a graph. The *line graph* $L(\Gamma_0)$ of Γ_0 is defined to be the graph whose vertex set is the edge set of Γ_0, two vertices of $L(\Gamma_0)$ being adjacent if and only if the corresponding edges of Γ_0 have a vertex in common. As a simple exercise, check that the graphs $T(n)$ and $L_2(n)$ are the line graphs of the complete graph K_n on n vertices and the complete bipartite graph $K_{n,n}$ with two bipartite blocks of size n, respectively.

Fact 1. Γ is represented in A_n if and only if it is the line graph of a bipartite graph on $n+1$ vertices.

PROOF. The vertex set of Γ is a set S of vectors of the form $\mathbf{e}_i - \mathbf{e}_j$, $0 \leq i, j \leq n$, $i \neq j$. Since all inner products are non-negative, each basis vector $\mathbf{e}_i$ has the same sign in every member of S containing it. So, if we let Γ_0 have vertex set $\{0,\ldots,n\}$, and join

i to j whenever $\pm(\mathbf{e}_i - \mathbf{e}_j) \in S$, then Γ_0 is bipartite (the bipartition corresponding to the signs associated with the basis vectors), and $\Gamma = L(\Gamma_0)$. □

The cocktail party graph $CP(m)$ of order $2m$ is represented in D_{m+1} by the vectors $\mathbf{e}_1 \pm \mathbf{e}_i$ for $i = 2, \ldots, m+1$.

Let Γ_0 be a graph with vertex set $\{v_1, \ldots, v_r\}$, and let $m_1, \ldots, m_r$ be non-negative integers. The *generalized line graph* $L(\Gamma_0;\ m_1, \ldots, m_r)$ is the disjoint union of $L(\Gamma_0)$ and cocktail party graphs $CP(m_i)$, $i = 1, \ldots, r$, together with all edges between each vertex $\{v_i, v_j\}$ of $L(\Gamma_0)$ and the cocktail parties $CP(m_i)$ and $CP(m_j)$. An example $L(\Gamma_0; 3, 1, 0, 2)$ is shown in Fig. 3.1.

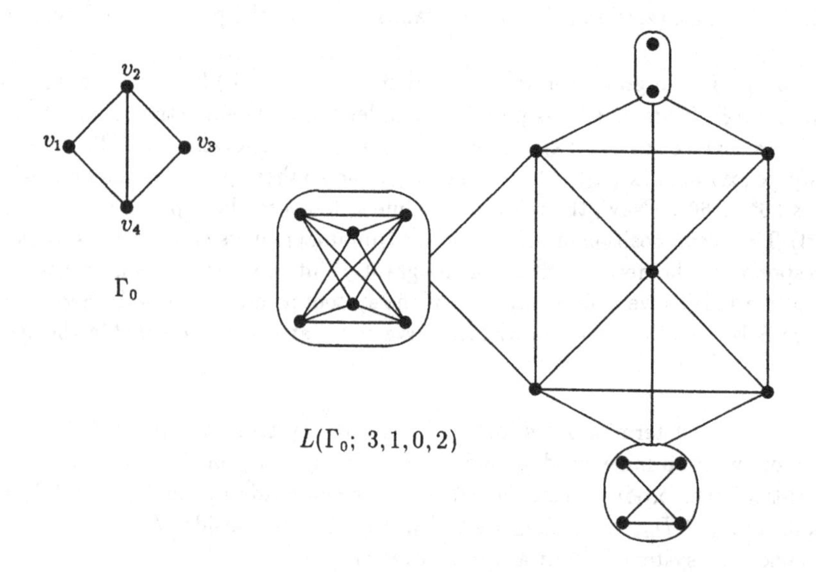

Fig. 3.1. A generalized line graph

If $n = \sum_{i=1}^{r}(1 + m_i)$, then the generalized line graph $L(\Gamma_0;\ m_1, \ldots, m_r)$ is represented in D_n, as follows. Number the basis vectors $\mathbf{e}_{i,j}$, for $i = 1, \ldots, r$ and $j = 0, \ldots, m_i$; then take

$$S = \{\mathbf{e}_{i,0} + \mathbf{e}_{j,0} : \{v_i, v_j\} \text{ an edge of } \Gamma_0\}$$
$$\cup \{\mathbf{e}_{i,0} \pm \mathbf{e}_{i,j} : j = 1, \ldots, m_i; i = 1, \ldots, r\}.$$

Fact 2. A graph is represented in D_n if and only if it is a generalized line graph.

PROOF. We have seen the reverse implication. So let Γ be a graph represented in D_n. Let S_0 be the set of basis vectors which always occur with the same sign in vertices of Γ. Without loss, we can assume that this sign is positive. Then the vectors

$$\{\mathbf{e}_i + \mathbf{e}_j \in S : \mathbf{e}_i, \mathbf{e}_j \in S_0\}$$

represent a line graph, say $L(\Gamma_0)$. Any basis vector $\mathbf{e}_i$ not in S_0 occurs in conjunction with at most one other basis vector $\mathbf{e}_j$, necessarily in S_0. (We cannot have vertices $\mathbf{e}_i \pm \mathbf{e}_j$ and $-\mathbf{e}_i \pm \mathbf{e}_k$ with $j \neq k$.) So we have a generalized line graph, in the representation just described.

Fact 3. Only finitely many graphs are represented in E_8. (Such graphs have at most 36 vertices; see Exercise 5.)

Summarizing:

(3.8) Theorem. *With finitely many exceptions, any connected graph whose least eigenvalue is -2 or greater is a generalized line graph. The exceptions are represented in E_8.* □

When can a connected generalized line graph be regular? We *claim* that this happens only if it is either a line graph or a cocktail party graph. For let $\{v_i, v_j\}$ be an edge of Γ_0, and let v_i, v_j have valencies k_i, k_j respectively. Then, in the generalized line graph $\Gamma = L(\Gamma_0;\ m_1, \ldots, m_r)$, the vertex $\{v_i, v_j\}$ has valency $(k_i - 1) + (k_j - 1) + 2m_i + 2m_j$, whereas a vertex in $CP(m_i)$ has valency $k_i + 2(m_i - 1)$. These numbers can never be equal; so either there are no cocktail parties (and $\Gamma = L(\Gamma_0)$), or Γ_0 has no edges and a single vertex v_1 (and $\Gamma = CP(m_1)$). We find the (unpublished) result of Hoffman and Ray-Chaudhuri:

(3.9) Theorem. *With finitely many exceptions, a connected regular graph with least eigenvlue -2 is either a line graph or a cocktail party graph. The exceptions are all represented in E_8.* □

Bussemaker, Cvetković and Seidel (1978) determined all the exceptional graphs. There are 187 regular connected graphs represented in E_8 which are not line graphs or cocktail party graphs, of which 68 are cospectral with line graphs.

From (3.9), we can determine all but finitely many strongly regular graphs with least eigenvalue -2.

(3.10) Theorem. With finitely many exceptions, a strongly regular graph with least eigenvalue -2 is isomorphic to $L_2(m)$, $T(m)$, or $CP(m)$ for some m.

PROOF. We have to show that, if $L(\Gamma_0)$ is strongly regular, then Γ_0 is K_m, $K_{m,m}$, or C_5, since $L(K_m) = T(m)$ and $L(K_{m,m}) = L_2(m)$, as we remarked earlier. (Note, incidentally, that $L(C_5) = C_5$ has least eigenvalue strictly greater than -2.)

So suppose that $L(\Gamma_0)$ is strongly regular.

Step 1. Using the constancy of k, we see that either Γ_0 is regular, or Γ_0 is bipartite and semiregular (that is, the valencies of vertices in a bipartite block are constant).

Step 2. Using the constancy of λ, we see that either Γ_0 is complete, or it contains no triangles; moreover, Γ_0 is regular (even if it is bipartite). We now *assume* that Γ_0 is not complete.

Step 3. The number μ is the (constant) number of edges joining two disjoint edges of Γ_0. Under our assumption, $\mu \leq 2$, with equality if and only if Γ_0 is complete bipartite. It is an easy exercise to show that, if $\mu = 1$, then $\Gamma_0 = C_5$. □

In the next chapter, we will finish the analysis by determining the exceptions in (3.10).

Special cases of the theorem can be proved by elementary combinatorial argument. Here is an example:

(3.11) Theorem. *A strongly regular graph Γ with the same parameters as $T(m)$, viz. $(\frac{1}{2}m(m-1), 2(m-2), m-2, 4)$, with $m > 8$, is isomorphic to $T(m)$.*

PROOF. For any vertex x of Γ, $\Gamma(x)$ carries a regular graph Δ of valency $m-2$ on $2(m-2)$ vertices. Let $y, z \in \Gamma(x)$ be non-adjacent, and let p vertices of $\Gamma(x)$ be adjacent to both. Since $\mu = 4$, and x is adjacent to y and z, we have $p \leq 3$. There are $m-2-p$ vertices of $\Gamma(x)$ adjacent to y but not z, $m-2-p$ adjacent to z but not y, and hence $p-2$ adjacent to neither; so $p \geq 2$. If $p = 3$, let w be the unique vertex of $\Gamma(x)$ adjacent to neither y nor z. Then every vertex of $\Gamma(x)$ adjacent to w is adjacent to either y or z, whence $m-2 \leq 3+3$, contrary to assumption. So $p = 2$.

Consider the complementary graph $\overline{\Delta}$. Suppose that it contains a circuit of odd length; choose such a circuit $C = (x_0, x_1, \ldots, x_{k-1}, x_k = x_0)$ with k minimal. There are no edges $\{x_i, x_j\}$ of $\overline{\Delta}$ with $i - j \not\equiv \pm 1 \pmod{k}$; for such an edge would divide C into two smaller circuits, at least one of which would have odd length. From the preceding paragraph, $k \neq 3$. Also, x_0 and x_1 are non-adjacent in Δ and have $k-4$ common neighbours $x_3, \ldots, x_{k-2}$ there; so $k \leq 6$. We must have $k = 5$. There are $m-5$ vertices adjacent to x_0 in $\overline{\Delta}$ (other than x_1 and x_4); these must be non-adjacent to both x_1 and x_4. Similarly, there are $m-5$ vertices non-adjacent to x_1 and x_3. Since x_1 is non-adjacent to $m-4$ vertices outside C, there are at least $m-6$ vertices outside C which are non-adjacent to both x_3 and x_4. By the preceding paragraph, $m-6 \leq 1$, a contradiction.

So $\overline{\Delta}$ contains no odd circuits, and is bipartite. This means that Δ contains two disjoint cliques (complete subgraphs) of size $m-2$. Equivalently, every vertex lies in two 'grand cliques' of size $m-1$, and any edge in a unique 'grand clique'. There are $\frac{1}{2}m(m-1) \cdot 2/(m-1) = m$ grand cliques, say $K_1, \ldots, K_m$. We can give the unique vertex in $K_i \cap K_j$ the label $\{i, j\}$. It is easily checked that this labelling identifies Γ with $T(m)$. □

Exercises

1. Prove that a graph with least eigenvalue -1 is a disjoint union of complete graphs. (An easy version of the geometric argument in this Chapter gives the most straightforward solution.)

2. Prove that the subsystem of E_8 consisting of lines orthogonal to two fixed perpendicular lines is isometric to D_6.

3. Prove that, if D_3 is defined analogously to D_n for $n \geq 4$, then D_3 is isometric to A_3.

4. Find a representation of the Petersen graph in E_8.

5. Use the argument in the proof of the absolute bound (2.23) to show that a graph which is represented in E_8 has at most 36 vertices. Construct such a graph with 36 vertices.

6. In the spirit of (3.11), show that a strongly regular graph with parameters $(m^2, 2(m-1), m-2, 2)$, $(m > 4)$, is isomorphic to $L_2(m)$.

7. Let S be one of the systems A_n, D_n, E_n, and let T be the set of lines making angles 0° or 45° with every line in S, and lying in the space spanned by S. Prove that

(a) if $S = A_n$ or $S = E_n$, then T is empty;
(b) if $S = D_n$ $(n > 4)$, then T consists of the lines spanned by the unit basis vectors $\mathbf{e}_i$ $(i = 1, \ldots, n)$;
(c) if $S = D_4$, then T is isometric to S.

8. Use the result of Exercise 7 to show the following:

If generalized line graphs $L(\Gamma;\ m_1, \ldots, m_r)$ and $L(\Delta;\ p_1, \ldots, p_s)$ are isomorphic, and $\sum_{i=0}^{r}(1 + m_i) > 4$, then $r = s$ and there is an isomorphism $\theta : \Gamma \to \Delta$ such that, if $\theta(v_i) = w_j$, then $m_i = p_j$.

(*Hint*: Suppose that either not all m_i are zero, or Γ is not bipartite. Show that the isomorphism induces an isometry of line systems. Now, by Exercise 7, the line system determines the unit basis uniquely up to sign change. Now reconstruct Γ and the integers m_i as in the proof of (3.8). For the line graph of a bipartite graph, a separate argument is required.)

REMARKS. 1. The result of this exercise is due to Cameron (1980b); it extends a theorem of Whitney (1932) on line graphs.

2. The theorem is false when $\sum_{i=0}^{r}(1+m_i) = 4$; can you give a counterexample?

4. Regular two-graphs

A regular graph can be regarded as a design whose points and blocks are its vertices and edges. It is a 1-$(n, 2, k)$ design if it has n vertices and its valency is k. Which regular graphs can be extended to 2-$(n+1, 3, k)$ designs?

We will not discuss the problem in such generality. Instead, we introduce the simplifying assumption that the graph uniquely determines its extension. In other words, the blocks (or triples) not containing the added point ∞ are determined by those containing ∞.

The simplest way in which this can be done is to assume that the number of 3-subsets of a 4-set which are blocks of the design is restricted to lie in a subset S of $\{0, \ldots, 4\}$. If any two members of S differ by at least 2, then membership in $\mathcal{B}$ of three 3-subsets of a 4-set determines membership of the fourth.

In detail, we have the following result.

(4.1) Proposition. *Let $\mathcal{B}$ be a collection of 3-subsets of X, where $|X| > 4$, not empty and not containing all 3-sets. Suppose that there is a subset S of $\{0, \ldots, 4\}$ such that*

(i) for any 4-subset Y of X, the number of members of $\mathcal{B}$ contained in Y belongs to S;

(ii) for $i, j \in S$, $i \neq j$, we have $|i - j| \geq 2$.

Then one of the following holds:

(a) $S = \{0, 3\}$, and $\mathcal{B}$ consists of all 3-subsets containing a fixed point of X;

(b) complement of (a);

(c) $S \subseteq \{0, 2, 4\}$.

PROOF. S is contained in a maximal set satisfying (ii), necessarily $\{1, 3\}$, $\{0, 3\}$, $\{1, 4\}$, or $\{0, 2, 4\}$. We have to show that the first is impossible while the second and third give conclusions (a) and (b) respectively.

A parity argument shows the impossibility of $\{1, 3\}$. Let Z be a 5-set, and count pairs (T, Y) with $T \subset Y \subset Z$, $T \in \mathcal{B}$, $|Y| = 4$. For each T, there are two choices of

Y, so the number of choices is even; but, for each of the 5 possible Y, the number of T is odd, so the total is odd.

Suppose that $S = \{0, 3\}$. Let $Y \subseteq X$ be maximal with respect to containing no member of $\mathcal{B}$. We can assume that $|Y| \geq 3$. If $x \notin Y$, then $\{a, b, x\} \in \mathcal{B}$ for some $a, b \in Y$. If $c \in Y$, consideration of $\{a, b, c, x\}$ shows that also $\{a, c, x\}, \{b, c, x\} \in \mathcal{B}$; then, by connectedness, $\{d, e, x\} \in \mathcal{B}$ for all $d, e \in Y$. Suppose that there is another point u outside Y. Then for any $a, b \in Y$, exactly one of $\{a, x, u\}, \{b, x, u\}$ belongs to $\mathcal{B}$. Considering three points $a, b, c \in Y$, we reach a contradiction. So x is the unique point outside Y, and $\mathcal{B}$ consists of all triples containing x.

The case $S = \{1, 4\}$ is obviously complementary to this case. □

We are left with one interesting case.

(4.2) DEFINITION. A *two-graph* is a collection $\mathcal{B}$ of 3-subsets of a set X, with the property that, for any 4-subset Y of X, an even number of members of $\mathcal{B}$ belong to Y.

The concept was introduced by G. Higman to study a 2-transitive action of the sporadic simple group Co_3; in the hands of Seidel, Taylor, and others, it has been used in connection with equiangular line systems, coverings of graphs, presentations of groups, and infinite homogeneous structures.

(4.3) DEFINITION. A two-graph is *regular* if it is a 2-design (with parameters 2-$(n, 3, \lambda)$ for some λ). The complete two-graph (where $\mathcal{B}$ consists of all 3-subsets) and the null two-graph (where $\mathcal{B}$ is empty) are regular. We call these the *trivial* two-graphs, and usually exclude them from the discussion.

The motivation for the term 'two-graph' comes from topology, where a 3-set or triangle is a 2-dimensional object. It should not be confused with the term 'k-graph', used in hypergraph theory to mean a k-uniform hypergraph, or set of k-subsets of a set. To keep the distinction clear, we write the word 'two' in full instead of using the numeral.

Given any graph Γ with vertex set X, let $\mathcal{B}$ be the set of 3-subsets of X which contain an odd number of edges of Γ. Then $(X, \mathcal{B})$ is a two-graph. (The proof of this is a parity argument similar to the one used in the proof of (4.1).) We say that Γ *yields* the two-graph $(X, \mathcal{B})$.

Every two-graph arises from this construction. For let $(X, \mathcal{B})$ be a two-graph, and x a point of X. Form a graph Γ on X by joining y to z whenever $\{x, y, z\} \in \mathcal{B}$. (Note that x is an isolated vertex.) Clearly the construction gives the correct triples

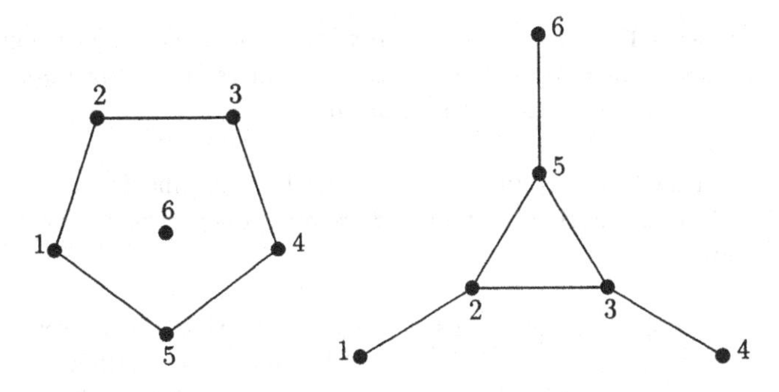

Fig. 4.1. Switching

containing x. For any 3-subset $\{y,z,w\}$ of $X \setminus \{x\}$,

$$\{y,z,w\} \in \mathcal{B} \Leftrightarrow \text{ an odd number of } \{x,y,z\},\{x,y,w\},\{x,z,w\} \in \mathcal{B}$$
$$\Leftrightarrow \text{ an odd number of } \{y,z\},\{y,w\},\{z,w\} \text{ are edges of } \Gamma,$$

so the rule works for these triples also.

In design terminology, every graph Γ can be uniquely extended to a two-graph, by adding an isolated vertex and applying the above construction. Where there is no ambiguity, we refer to this two-graph as the *extension* of Γ.

Different graphs on X can yield the same two-graph. We now investigate this.

(4.4) DEFINITION. The operation of *switching* a graph Γ with respect to a set Y of vertices replaces Γ by the graph Γ' such that $\{x,y\}$ is an edge of Γ' if and only if both or neither of the following statements hold:
(a) $|\{x,y\} \cap Y| = 0$ or 2;
(b) $\{x,y\}$ is an edge of Γ.
In other words, switching replaces all edges between Y and its complement with non-edges and *vice versa*, leaving edges within Y and outside Y unaltered. Note that switching with respect to Y and $X \setminus Y$ are the same operation. Furthermore, switching successively with respect to Y_1 and Y_2 is the same as switching with respect to the symmetric difference $Y_1 \Delta Y_2$. Thus the set of all graphs on the vertex set X falls into equivalence classes, called *switching classes*, each containing 2^{n-1} graphs (where $n = |X|$).

We saw an application of switching in Example (2.21). For a simple example, see Fig. 4.1: the second graph in that figure may be obtained from the first by switching with respect to the set $\{5\}$. It can be verified that the switching class of these graphs contains six graphs isomorphic to the first graph in Fig. 4.1, six copies of its complement, ten copies of the second graph, and ten of its complement: 32 in all, as expected.

(4.5) Theorem. *Graphs Γ_1 and Γ_2 on the vertex set X yield the same two-graph if and only if Γ_1 can be switched into Γ_2. In other words, there is a natural bijection between two-graphs and switching classes of graphs on X.*

PROOF. The fact that switching-equivalent graphs yield the same two-graph is a consequence of the observation that switching does not change the parity of the number of edges within a 3-set.

Conversely, suppose that Γ_1 and Γ_2 have the property that the parity of the number of edges within any triple is the same in each. Set $x \equiv y$ if either $x = y$ or the character of $\{x, y\}$ (edge or non-edge) is the same in each graph. By hypothesis, this is an equivalence relation; and it cannot have more than two equivalence classes, since if three vertices x, y, z lie in different classes then the triple $\{x, y, z\}$ has different parity in the two graphs. Now switching with respect to an equivalence class of $\equiv$ transforms Γ_1 into Γ_2. □

Some new strongly regular graphs can be obtained by switching. We saw an example of this in (2.21).

(4.6) EXAMPLE. The *Shrikhande graph* (Shrikhande (1959)). This graph is obtained from $L_2(4)$ by switching with respect to the diagonal set of vertices, viz. $\{(1,1), (2,2), (3,3), (4,4)\}$. It is strongly regular with the same parameters as $L_2(4)$, viz. (16, 6, 2, 2). It has another nice representation as a graph drawn on a torus, as shown in Fig. 4.2, where opposite sides are identified according to the arrows.

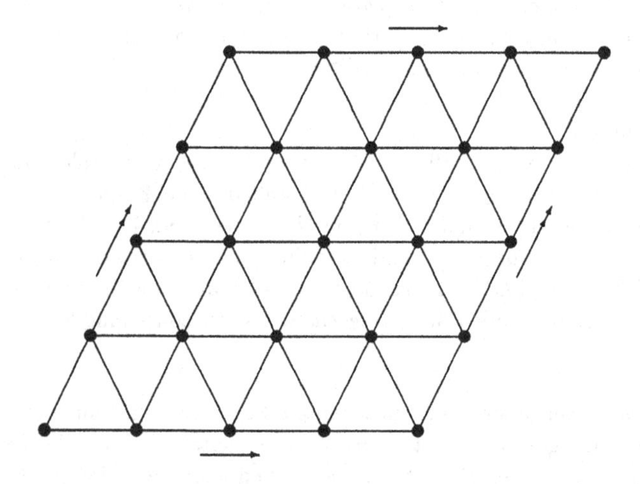

Fig. 4.2. The Shrikhande graph

(4.7) EXAMPLE. The *Chang graphs* (Chang (1960)). The vertex set of $T(8)$ is the set of 2-subsets of $\{1, \ldots, 8\}$. So a subset of the vertex set can be regarded as the edge

set of a graph with 8 vertices. The three Chang graphs are obtained by switching $T(8)$ with respect to
(1) four disjoint edges;
(2) an octagon;
(3) the disjoint union of a pentagon and a triangle.
Each is strongly regular, with the same parameters as $T(8)$, viz. (28, 12, 6, 4).

(4.8) REMARK. Switching $T(8)$ with respect to the disjoint union of two K_4s produces a graph isomorphic to $T(8)$. Similarly, switching $L_2(4)$ with respect to the 8-set

$$\{(1,1),\ (1,2),\ (2,1),\ (2,2),\ (3,3),\ (3,4),\ (4,3),\ (4,4)\}$$

gives a graph isomorphic to $L_2(4)$.

In connection with switching, it is convenient to use a slightly modified adjacency matrix for a graph Γ. Let B be the $n \times n$ matrix whose (i,j) entry is

$$b_{ij} = \begin{cases} 0 & \text{if } i = j, \\ -1 & \text{if } \{x_i, x_j\} \text{ is an edge,} \\ +1 & \text{if } \{x_i, x_j\} \text{ is a non-edge,} \end{cases}$$

where $X = \{x_1, \ldots, x_n\}$. If Γ yields the two-graph $(X, \mathcal{B})$, then $\{x_i, x_j, x_k\} \in \mathcal{B}$ if and only if $b_{ij}b_{jk}b_{ki} = -1$. Also, if Γ' is obtained by switching Γ with respect to Y, then Γ' has adjacency matrix $B' = DBD$, where $D = (d_{ij})$ is a diagonal matrix with

$$d_{ii} = \begin{cases} -1 & \text{if } x_i \in Y, \\ +1 & \text{if } x_i \notin Y. \end{cases}$$

Thus B and B' are similar matrices, and have the same spectrum. So the spectrum is an invariant of the switching class, and so (by (4.5)) of the two-graph. We speak of the eigenvalues of a two-graph in this sense.

(4.9) Theorem. *A two-graph is regular if and only if it has just two distinct eigenvalues.*

PROOF. Let Γ be a graph yielding $(X, \mathcal{B})$, and B its adjacency matrix. Since B is symmetric and has $n-1$ entries ± 1 and one entry 0 in each row, we see that B^2 has constant diagonal $n-1$. Now B has two eigenvalues if and only if it satistfies a quadratic equation, whoch must be of the form $B^2 = \alpha B + (n-1)I$ for some integer α; this holds if and only if, for $i \neq j$, the (i,j) entry $(B^2)_{ij}$ of B^2 is equal to αb_{ij}.

Now $(B^2)_{ij} = \sum_{k \neq i,j} b_{ik}b_{kj}$; and $b_{ik}b_{kj}$ is equal to $-b_{ij}$ if $\{x_i, x_j, x_k\} \in \mathcal{B}$, or b_{ij} otherwise. So, if x_i and x_j are contained in λ_{ij} members of $\mathcal{B}$, then

$$(B^2)_{ij} = -\lambda_{ij}b_{ij} + (n-2-\lambda_{ij})b_{ij} = (n-2-2\lambda_{ij})b_{ij}.$$

Thus, B has two eigenvalues if and only if λ_{ij} is constant. □

(4.10) EXAMPLE. There is a unique 2-(6, 3, 2) design $\mathcal{D}$ (Exercise 1). It has the property that exactly one of each complementary pair of 3-sets is a block. Hence, given a 4-set Y, two blocks contain the complement of Y, and so two blocks are contained in Y. Thus $\mathcal{D}$ is a regular two-graph.

The corresponding switching class contains the graphs of Fig. 4.1. The adjacency matrix B of one of these graphs satisfies $B^2 = 5I$, and so has eigenvalues $\pm\sqrt{5}$, each with multiplicity 3.

There is another characterization of regular two-graphs. Let Γ_x denote the unique graph in a given switching class for which x is an isolated vertex (i.e. for which the two-graph is the extension obtained by adjoining x). For the regular two-graph of (4.10), Γ_x is the pentagon (for any vertex x).

(4.11) Theorem. *For a non-trivial two-graph $(X, \mathcal{B})$, the following are equivalent:*
(a) $(X, \mathcal{B})$ is regular;
(b) for all $x \in X$, Γ_x is strongly regular with $k = 2\mu$;
(c) for some $x \in X$, Γ_x is strongly regular with $k = 2\mu$.

PROOF. Let k_y be the valency of y, and γ_{yz} the number of common neighbours of y and z, in the graph Γ_x. Then the number of triples containing x and y is k_y, and the number containing y and z is $k_y + k_z - 2\gamma_{yz}$ if y and z are non-adjacent, or $n - k_y - k_z + 2\gamma_{yz}$ if y and z are non-adjacent. So $(X, \mathcal{B})$ is regular if and only if Γ_x is strongly regular with parameters satisfying

$$k = 2(k - \mu) = n - 2(k - \lambda).$$

Now the equation

$$k(k - \lambda - 1) = (n - 2 - k)\mu$$

shows that, if $k = 2\mu$, then $n - 2(k - \lambda) = k$; so the second inequality is a consequence of the first. □

(4.12) Corollary. *A non-trivial regular two-graph has an even number of points.*

PROOF. $n = 3k - 2\lambda$, and $k = 2\mu$ is even. □

There is another link between strongly regular graphs and regular two-graphs.

(4.13) Theorem. *Let Γ be a regular graph which yields the two-graph $(X, \mathcal{B})$. Then $(X, \mathcal{B})$ is a regular two-graph if and only if Γ is strongly regular with eigenvalues k, r, s satisfying $n = 2(k - r)$ or $n = 2(k - s)$.*

PROOF. If A is the usual adjacency matrix of Γ, then the modified $(0, \mp 1)$ adjacency matrix B is given by $B = J - I - 2A$. Because Γ is regular, A commutes with J, and so A and B have the same eigenvectors.

If the two-graph is strongly regular, then B has only two distinct eigenvalues, and so $A|\mathbf{j}^{\perp}$ has only two distinct eigenvalues; so Γ is strongly regular, by (2.19).

Conversely, if Γ is strongly regular, then $\mathbf{j}$ is an eigenvector of B with eigenvalue $n-1-2k$, and the other eigenvalues of B are $-1-2r$ and $-1-2s$. So by (4.11), $(X,\mathcal{B})$ is regular if and only if either $n-1-2k=-1-2r$ or $n-1-2k=-1-2s$.□

We now have all the machinery necessary to give Seidel's proof of the classification of strongly regular graphs with least eigenvalue -2. See Shrikhande (1959), Hoffman (1960), Chang (1959, 1960), Seidel (1967, 1968, 1974) for the development of this theorem.

(4.14) Theorem. *A strongly regular graph with least eigenvalue -2 is one of the following:*
(a) $T(m)$, $m \geq 5$;
(b) $L_2(m)$, $m \geq 3$;
(c) $CP(m)$, $m \geq 2$;
(d) the Petersen graph;
(e) the complement of the Clebsch graph;
(f) the complement of the Schläfli graph;
(g) the Shrikhande graph;
(h) the three Chang graphs.

PROOF. By (3.10), either we have (a)–(c), or our graph Γ is represented in E_8. In the latter case, $A+2I$ has rank at most 8, and so the eigenvalue r of A has multiplicity at most 7. By (2.23), Γ has at most 35 vertices. Now it is straightforward to list all the possible parameters satisfying the integrality and Krein conditions and in which $s=-2$ and r has multiplicity 5, 6 or 7. (One parameter set, viz. (28, 9, 0, 4), requires the Krein condition, or the result of Exercise 2 of Chapter 2.) The resulting list consists of the parameters of (d)–(g) in the theorem and of $T(6)$.

For the parameter sets (15, 8, 4, 4) and (27, 16, 10, 8), the complementary graph has parameters (15, 6, 1, 3) or (27, 10, 1, 5) respectively. Thus our 'recurring theme' reappears; by (2.4), these are $T(6)$ and the complement of the Schläfli graph respectively.

In the remaining cases, with 10, 16 or 28 vertices, it happens that all possible parameter sets satisfy the conditions of (4.13), and so the unknown graph yields a regular two-graph $(X,\mathcal{B})$. By (4.11), the derived graph is strongly regular with parameters (9, 4, 1, 2), (15, 6, 1, 3), or (27, 10, 1, 5) respectively. Again, by (2.4), we know these graphs! We conclude that, for each of the three numbers of vertices, the unknown strongly regular graphs fall into a single switching class, and thus are obtained from known ones by switching. We have to investigate how this switching can be done. We give the argument for $n=28$; the other cases are considerably simpler.

Our problem, then, is to find all switching sets Y of vertices which switch $T(8)$ into a regular graph with valency 12. As in (4.7), the switching set is the edge set of a graph Δ on 8 vertices, and a short argument shows that this graph is regular. Moreover, we can assume that Y is smaller than its complement, from which it follows that Δ has valency at most 3.

Before proceeding further, we elaborate on (4.7), where we observed that the graph obtained by switching $T(8)$ with respect to the edge set of $2K_4$ (the disjoint union of two K_4s) is isomorphic to $T(8)$. If the vertex sets of the K_4s are $\{1,\ldots,4\}$ and $\{5,\ldots,8\}$, then an isomorphism between the unswitched and switched graphs is given by the following permutation:

- for each $\{i,j\} \subset \{1,\ldots,4\}$, interchange $\{i,j\}$ with $\{1,\ldots,4\} \setminus \{i,j\}$;
- for each $\{i,j\} \subset \{5,\ldots,8\}$, interchange $\{i,j\}$ with $\{5,\ldots,8\} \setminus \{i,j\}$;
- fix all other 2-sets.

We will call this process *reflection*.

The upshot is that, if Δ and Δ' are regular graphs on $\{1,\ldots,8\}$, and if Δ' is obtained from Δ by taking the symmetric difference of its edge set with $2K_4$ followed by reflection, then the graphs obtained by switching $T(8)$ with respect to Δ and Δ' are isomorphic. We call this the *SDR* (= 'symmetric difference plus reflection') *process*.

In particular, $2K_4$ is equivalent to the null graph under SDR. Also, $2C_4$ (two disjoint 4-cycles) is equivalent to $4K_2$ (four disjoint edges). Fig. 4.3 shows a less trivial instance of SDR, where the complete graphs have vertex sets $\{1,2,5,8\}$ and $\{3,4,6,7\}$ respectively.

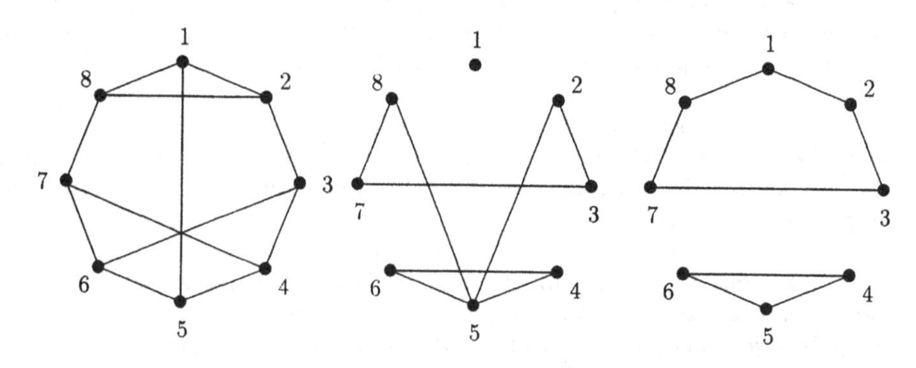

Fig. 4.3. Symmetric difference and reflection

Now we examine regular graphs of valency m on 8 vertices, for $m = 1, 2, 3$.

Case $m = 1$: The only graph of valency 1 is $4K_2$, which gives rise to Chang (1).

Case $m = 2$: The three graphs are C_8, $C_5 + C_3$, and $2C_4$. The first two give us Chang (2) and Chang (3); the third gives Chang (1) again, by SDR.

Case $m = 3$: There are six graphs of valency 3 on 8 vertices. Use of SDR shows that no new strongly regular graphs arise. (We have seen this in two cases. The reader is encouraged to try the others.)

Finally, note that the Chang graphs are not isomorphic to $T(8)$ (since none contains a 7-clique), nor to one another (they contain, respectively, 8, 0 and 3 6-cliques). □

(4.15) EXAMPLE. Next we give a construction of Higman's regular two-graph on 276 points. Let $(Y, \mathcal{B})$ be the unique 4-(23, 7, 1) design, with 253 blocks. Recall that any two blocks of this design meet in either 1 or 3 points. Take $X = Y \cup \mathcal{B}$, and let Γ be the graph with vertex set X defined by the following rules.
(a) any two vertices in Y are adjacent;
(b) a vertex in Y and a vertex in $\mathcal{B}$ are adjacent if and only if they are incident;
(c) two vertices in $\mathcal{B}$ are adjacent if and only if their intersection has cardinality 3.
Then Γ yields the required two-graph.

Next, we discuss a theorem of Neumaier (1982a), which has a strong design-theoretic flavour. A *clique* in a two-graph $(X, \mathcal{B})$ is a subset Y of X, all of whose 3-subsets belong to $\mathcal{B}$. There is an easy upper bound for the size of a clique in terms of the spectrum.

(4.16) Proposition. *Let Y be a clique in a two-graph with smallest eigenvalue ρ. Then $|Y| \leq 1 - \rho$.*

PROOF. There is a graph in the switching class for which Y is a clique (in the graph-theoretic sense). Let B be its adjacency matrix. Then $B - \rho I$ is positive semi-definite, and so is its principal sub-matrix $B' - \rho I$ corresponding to the subgraph Y. But $B' = -J + I$; so $-J + (1 - \rho)I$ is positive semi-definite, whence $|Y| \leq 1 - \rho$. □

(4.17) DEFINITION. A two-graph $(X, \mathcal{B})$ is called *t-regular* if, for $2 \leq i \leq t$, the number a_i of $(i + 1)$-cliques containing a given i-clique depends only on i. (Note that '2-regular' is the same as 'regular', as defined earlier, and (4.11) shows that any 2-regular two-graph is 3-regular: the numbers a_2 and a_3 are the parameters k and λ of the derived strongly regular graph.) A two-graph is *completely regular* if it is t-regular for $t = -\rho$, where ρ is the smallest eigenvalue. (By (4.16), this is the largest value we need consider.)

(4.18) Theorem. *Let $(X, \mathcal{B})$ be a completely regular two-graph on n vertices, with eigenvalues ρ_1 and ρ_2. Then (n, ρ_1, ρ_2) is one of the following:*
(a) $(10, 3, -3)$;
(b) $(16, 5, -3)$;

(c) $(28, 9, -3)$;
(d) $(36, 7, -5)$;
(e) $(276, 55, -5)$;
(f) $(1128, 161, -7)$;
(g) $(3160, 351, -9)$.
The two-graphs in cases (a)–(e) are unique; existence is unknown in the remaining cases. □

REMARK. Neumaier also included possible examples with $n = 96$, $(\rho_1, \rho_2) = (19, -5)$, and with $n = 288$, $(\rho_1, \rho_2) = (41, -7)$. These were subsequently eliminated by Blokhuis and Brouwer (1984) and Blokhuis and Wilbrink (1989).

Note that cases (a)–(c) are the extensions of the three graphs of (2.3)(c). Case (e) is the complement of Higman's two-graph (4.14). Case (d) is the extension of the graph whose vertices are the 3-subsets of a 7-set, two vertices being adjacent if they intersect in an even number of points. (This graph has several descriptions. For example, it arises from a quadratic form over $\mathbb{F}_2$ by the construction of (2.29).)

Returning to the general theme of extensions of graphs, a number of interesting questions can be formulated. We mention just one.

Let $\mathcal{D} = (X, \mathcal{B})$ be a biplane (a square 2-$(v, k, 2)$ design, with $v = \frac{1}{2}(k^2 - k + 2)$). Let $\mathcal{B}'$ be the set of all triples which are contained in a block of $\mathcal{D}$. Then $\mathcal{D}' = (X, \mathcal{B}')$ is a 2-$(v, 3, 2(k-2))$ design, having the property that all its derived designs are isomorphic to $T(k)$. For, given x, if $B_1, \ldots, B_k$ are the blocks containing x, then any further point y is in just two of these blocks, say B_i and B_j; the map $y \mapsto \{i, j\}$ is a bijection, which induces an isomorphism to $T(k)$.

PROBLEM. Does every 2-$(v, 3, 2(k-2))$ design, all of whose derived designs are isomorphic to $T(k)$, arise from a biplane in this way?

Exercises

1. Show that there is a unique 2-(6, 3, 2) design, up to isomorphism. Show that it has the property that, out of each complementary pair of 3-sets, exactly one is a block. Deduce that the design is a regular two-graph.

2. Find a set of vertices which switches $T(5)$ into the Petersen graph.

3. Prove directly that the two-graph yielded by the regular graph Γ is itself regular if and only if Γ is strongly regular with parameters (n, k, λ, μ) which satisfy $n = 2(2k - \lambda - \mu)$.

4. Prove that the Shrikhande and Chang graphs are strongly regular.

5. Prove that Higman's two-graph is regular, with $\lambda = 162$. Deduce the existence of a strongly regular graph with parameters (275, 162, 105, 81). (The complement of this graph is *McLaughlin's graph*, constructed by McLaughlin (1969).)

6. Let Γ be strongly regular, with parameters (275, 162, 105, 81), and A its adjacency matrix. Let Γ_1 be the induced subgraph on $\Gamma(x)$, for some vertex x, and A_1 its adjacency matrix (a principal submatrix of A).

(a) Prove that $A + 3I - \frac{3}{5}J$ has rank 22. Observing that $\mathbf{j}$ is an eigenvector of $A_1 + 3I - \frac{3}{5}J$ with non-zero eigenvalue, prove that A_1 has eigenvalue -3 with multiplicity at least 140.

(b) Let $105, \alpha_1, \ldots, \alpha_{21}$ be the other eigenvalues of A_1. By computing the traces of A_1 and A_1^2, and using the variance trick, prove that $\alpha_1 = \ldots = \alpha_{21} = 15$.

(c) Deduce that Γ_1 is strongly regular, with parameters (162, 105, 72, 45).

7. For this exercise, you may use without proof the uniqueness of the 4-(23,7,1) design, and the fact that its automorphism group is transitive both on points and on blocks, as well as the result of Exercise 6.

Let $(X, \mathcal{B})$ be Higman's two-graph. Let Z be a set of 23 points, all of whose 3-subsets belong to $\mathcal{B}$. (The set Y in the construction has this property.)

For each point $x \notin Z$, define a relation $\equiv$ on Z by $z_1 \equiv z_2$ if and only if $\{x, z_1, z_2\} \in \mathcal{B}$. Prove that $\equiv$ is an equivalence relation, with two equivalence classes of sizes 7 and 16. [Hint: if $n_i, 23 - n_i$ are the sizes of the equivalence classes for $i = 1, \ldots, 253$, calculate $\sum(n_i - 7)^2(n_i - 16)^2$. In fact it is enough to show that the value of this sum can be calculated, since it is clearly zero in the case where $Z = Y$.]

Prove that, if $\mathcal{C}$ is the set of classes of size 7, for each $x \notin Z$, then $(Z, \mathcal{C})$ is a 4-(23, 7, 1) design, and that the two-graph constructed from $(Z, \mathcal{C})$ by Higman's construction is precisely $(X, \mathcal{B})$ (if we identify points outside Z with blocks of the design).

By finding a set Z different from Y, deduce that Higman's two-graph admits a transitive automorphism group.

(This group is the sporadic simple group Co_3 of Conway (1969).)

8. Prove the assertion in the text that a set Y which switches $T(8)$ into a regular graph of the same valency must be the edge set of a regular graph on 8 vertices.

9. Show that $L_2(4)$, the Shrikhande graph, and the Clebsch graph are associated with different polarities of the same 2-(16, 6, 2) design.

5. Quasi-symmetric designs

Recall that a square or 'symmetric' design has the property that any two blocks intersect in a constant number of points.

(5.1) DEFINITION. A 2-design is called *quasi-symmetric* if the number of points in the intersection of two blocks takes only two values.

In particular, any affine design is quasi-symmetric.

In this chapter we consider such designs. We prove a theorem of Goethals and Seidel, showing that quasi-symmetric designs are very closely connected with strongly regular graphs. In the remainder of the chapter, we give some examples and some characterizations, and look at a few results on a related class of designs (square designs in which the number of blocks containing three points takes only two values).

Throughout this section, there will be designs whose parameters k and λ are very different from the parameters with the same name of the strongly regular graphs associated with them. To avoid confusion, we use the symbols (n, a, c, d) for the parameters of a strongly regular graph.

In a 2-$(v, k, 1)$ design which is not a projective plane, two blocks intersect in 0 or 1 point. The *line graph* of the design is the graph whose vertices are the blocks (or 'lines'), two vertices being adjacent precisely when they intersect in a point. This generalizes the triangular graph $T(n)$ (the case $k = 2, v = n$). The line graphs of 2-designs with $\lambda = 1$ were studied by Bose (1963), who made the following observation.

(5.2) Proposition. *The line graph of a 2-$(v, k, 1)$ design with $b > v$ is strongly regular.*

Indeed, easy counting arguments give us the parameters:

$$n = b = \frac{v(v-1)}{k(k-1)},$$

$$a = k(r-1) = \frac{k(v-k)}{(k-1)},$$

$$c = (r-2) + (k-1)^2 = \frac{v-2k+1}{k-1} + (k-1)^2,$$
$$d = k^2.$$

The graph is not the complete graph, since $b > v$, and is not the null graph, since $r > 1$. □

Perhaps not surprisingly, the integrality condition gives us no non-existence criteria for such designs beyond the fact that the parameters b and r are integers. We will soon see the real reason for this, in a more general setting.

Let x and y, where $x < y$, be the two cardinalities of block intersections in the quasi-symmetric design $\mathcal{D}$. The *block graph* of $\mathcal{D}$ has as vertices the blocks of $\mathcal{D}$, two vertices adjacent if they intersect in y points. For example, the block graph of an affine design is complete multipartite. (The converse is false; see Exercise 6.) Note that $\mathcal{D}$ and its complement have the same block graph.

Goethals and Seidel (1970) generalized Bose's observation as follows.

(5.3) Theorem. *The block graph of a quasi-symmetric 2-design is strongly regular.*

PROOF. We prove this by matrix methods resembling those used for Fisher's inequality (1.12). Let M be the incidence matrix of the design. Then $M^\top M = (r-\lambda)I + \lambda J$ has eigenvalues $r - \lambda + \lambda v = rk$ with multiplicity 1 (corresponding to the all-1 eigenvector) and $r - \lambda$ with multiplicity $v - 1$. Now $MM^\top$ has the same non-zero eigenvalues as $M^\top M$ with the same multiplicities (and in particular, the eigenvector for the eigenvalue rk is also $\mathbf{j}$), but has in addition the eigenvalue 0 with multiplicity $b - v$.

On the other hand, we have

$$\begin{aligned} MM^\top &= kI + yA + x(J - I - A) \\ &= (k-x)I + (y-x)A + xJ, \end{aligned}$$

where A is the adjacency matrix of the block graph. So, on the space $\mathbf{j}^\perp$, the matrix A has just the two eigenvalues $(r - \lambda - k + x)/(y - x)$ with multiplicity $v - 1$, and $-(k - x)/(y - x)$ with multiplicity $b - v$. By (2.19), the block graph is strongly regular. □

REMARK. The valency a of the strongly regular graph can be found by applying the above equation to $\mathbf{j}$, giving

$$rk = (k-x) + (y-x)a + xb,$$

or $a = ((r-1)k - xb - y)/(y - x)$. The other parameters can be calculated from the eigenvalues by the formulae following (2.19).

(5.4) Corollary. *In a quasi-symmetric 2-design, $y - x$ divides $k - x$ and $r - \lambda$.* □

A proof of (5.3) by elementary counting was suggested to the authors by C. W. Norman, and is outlined in Exercise 1.

In their paper, Goethals and Seidel went on to examine various familiar strongly regular graphs, to see which could occur as block graphs of quasi-symmetric 2-designs. We summarize their results.

(5.5) Theorem. *Let $\mathcal{D}$ be a quasi-symmetric 2-design with block graph Γ.*
(a) Γ cannot be a ladder graph (but, if repeated blocks are allowed, then if Γ is a ladder graph then $\mathcal{D}$ is a square design with each block repeated twice).
(b) If Γ is a cocktail party graph, then $\mathcal{D}$ is a Hadamard 3-design.
(c) Γ is never a square lattice graph $L_2(n)$ or the complement of one. □

We will give a proof of (b) later. We turn next to some extremal results for quasi-symmetric designs.

(5.6) Proposition. *Let $\mathcal{D}$ be a quasi-symmetric 2-design with block intersection numbers x and y.*
(a) $b \leq v(v-1)/2$, with equality if and only if $\mathcal{D}$ is a 4-design.
(b) If $x = 0$, then $b \leq v(v-1)/k$, with equality if and only if $\mathcal{D}$ is a 3-design.

PROOF. Part (a) is a special case of (1.41).

For (b), suppose that $x = 0$, and let $\mathcal{E} = \mathcal{D}_p^\top$, the design whose points and blocks are the blocks of $\mathcal{D}$ containing p and the points of $\mathcal{D}$ different from p respectively, where p is a point of $\mathcal{D}$. By assumption, $\mathcal{E}$ is a 2-$(r, \lambda, y-1)$ design, having $v-1$ blocks. By Fisher's inequality, $v - 1 \geq r$ (or equivalently, $b \leq v(v-1)/k$), with equality if and only if $\mathcal{D}_p$ is a 2-design, that is, $\mathcal{D}$ is a 3-design. (Note that, if the bound is met, then $\mathcal{D}$ is an extension of a square 2-design, and (1.28) applies.) □

There can exist no (non-trivial) square 3-design or quasi-symmetric 5-design, in consequence of the following result.

(5.7) Proposition. *A tight $2s$-design with $k < v - s$ cannot be a $2s+1$-design.*

PROOF. If $\mathcal{D}$ is such a design, then $\mathcal{D}_p$ is also tight; so $b = \binom{v}{s}$, $r = \binom{v-1}{s}$. The equation $vr = bk$ then shows that $k = v - s$. □

In addition, (1.43) shows that, up to complementation, there is a unique quasi-symmetric 4-design, the 4-(23, 7, 1) design $\mathcal{D}$. Its derived and point-residual designs are quasi-symmetric 3-designs, and their derived and point-residual designs are quasi-symmetric 2-designs (except for the second derived design of $\mathcal{D}$, which is square, being the projective plane of order 4).

These designs have strongly regular block graphs, most of which have not appeared yet in the earlier chapters. The parameters of the graphs are given in Table 5.1

Design	n	a	c	d	r	s
$\mathcal{D}$	253	112	36	60	2	−26
$\mathcal{D}_p$	77	16	0	4	2	−6
$\mathcal{D}^p$	176	70	18	34	2	−18
$\mathcal{D}_p^q$	56	10	0	2	2	−4
$\mathcal{D}^{pq}$	120	42	8	18	2	−12

Table 5.1.

REMARKS. (1) We have tabulated the parameters of the complements of the block graphs, since these are smaller (and the graphs are more familiar).

(2) The blocks of $\mathcal{D}_p^q$, resp. $\mathcal{D}^{pq}$, form a class of ovals, resp. Baer subplanes, in PG(2,4), as in Lüneburg's construction described in Chapter 1.

(3) The block graph of $\mathcal{D}$, and the graph $T(23)$, have the same multiplicities of eigenvalues, and yet their parameters are different.

(4) The design $\mathcal{D}_p$ attains the bound of (5.6)(b), since it is an extension of a square design.

(5) The complement of the block graph of $\mathcal{D}_p^q$ is the Gewirtz graph.

We now give a result which implies (5.5)(b).

(5.8) Theorem. *A quasi-symmetric design with $b = 2v - 2$ is either a Hadamard design or the unique 2-(6, 3, 2) design.*

(The eigenvalues of the cocktail party graph $CP(n)$ have multiplicities 1, n, $n-1$; so, if it is the block graph of $\mathcal{D}$, then we have $v = n+1$, $b - v = n - 1$, whence $b = 2v - 2$.)

PROOF. From the hypothesis, we draw two conclusions:

(a) Since $bk = vr$, we must have $v = 2k$, $b = 4k - 2$.

(b) The eigenvalues of the block graph have multiplicities 1, $v-1$, $b - v = v - 2$; so Wielandt's theorem (2.20) applies.

Given a block, count the number of choices of a point in that block and another block through that point:

$$ay + (4k - 3 - a)x = 2k(k-1).$$

If the graph or its complement is a ladder graph, then we can assume (possibly exchanging x and y) that $a = n - 2 = 4(k-1)$, and so

$$4(k-1)y + x = 2k(k-1).$$

Thus, $2(k-1)$ divides x. Since $x < k$, we have $x = 0$, $y = k/2$. By (5.5)(b), $\mathcal{D}$ is a 3-design which is an extension of a square 2-design, and so it is a Hadamard 3-design.

In the other case in Wielandt's Theorem, we have $v - 1 = 2s^2 + 2s + 1$, $k = s^2 + s + 1$, $a = s(2s+1)$, and so

$$s(2s+1)y + (s+1)(2s+1)x = 2s(s+1)(s^2+s+1).$$

Thus $2s+1$ divides $2s(s+1)(s^2+s+1)$, which implies that $s = 1$. Then $\mathcal{D}$ is a 2-(6, 3, 2) design, which is unique (Exercise 1 of Chapter 4). □

A situation in which knowledge of the block graph of a design is useful is the *Hall–Connor theorem*, promised after (1.40):

(5.9) Theorem. *A quasi-residual design with $\lambda = 2$ is residual.*

PROOF. Recall that $\mathcal{D}$ is quasi-residual if $v\lambda = k(k + \lambda - 1)$, or (in our case) $v = \frac{1}{2}k(k+1)$. Let B be a block of $\mathcal{D}$, and let n_i denote the number of blocks which intersect B in i points ($i \geq 0$). As in (1.15), we find by counting choices of a block $B' \neq B$ and j points in $B \cap B'$, $j = 0, 1, 2$:

$$\sum n_i = \tfrac{1}{2}k(k+3),$$
$$\sum in_i = k(k+1),$$
$$\sum i(i-1)n_i = k(k-1).$$

Thus

$$\sum (i-1)(i-2)n_i = 0,$$

whence $n_i = 0$ unless $i = 1$ or 2. So $\mathcal{D}$ is quasi-symmetric, with $x = 1$, $y = 2$.

Let Γ be the complement of the block graph of $\mathcal{D}$; that is, blocks are adjacent if and only if they intersect in 1 point. A short calculation establishes that the strongly regular graph Γ has the same parameters as $T(k+2)$.

Suppose first that $k \neq 6$. By (3.11), $\Gamma \cong T(k+2)$. Thus, there is a one-to-one correspondence between blocks of D and 2-subsets of a $(k+2)$-set C, so that blocks meet in 1 or 2 points according as the corresponding 2-sets meet in 1 or 0 points. Thus, if we adjoin the elements of C to the point set, adding to each block the corresponding pair of elements of C, and let C be a new block ('at infinity'), we obtain a square 2-$(v+k+2, k+2, 2)$ design $\mathcal{E}$, such that $\mathcal{E}^C = \mathcal{D}$.

In the case $k = 6$, the square design of which $\mathcal{D}$ would be the residual would be a 2-(29, 8, 2) design, which does not exist (by the Bruck–Ryser–Chowla theorem

(1.16)). So we have to show that no 2-(21, 6, 2) design exists either. As before, we know that the design is quasi-symmetric; if the complement of its block graph is $T(8)$, then the same argument as before applies. So we may assume that the complement of the block graph is a Chang graph. *Ad hoc* arguments complete the proof. □

Using an elaboration of these ideas, Bose, Shrikhande and Singhi (1977) proved the following result.

(5.10) Theorem. *There is a function g defined on the natural numbers with the property that a quasi-residual design with $k > g(\lambda)$ is residual.* □

The function they gave was

$$g(\lambda) = \begin{cases} 76 & \text{for } \lambda = 3, \\ \frac{1}{2}(\lambda-1)(\lambda^4 - 2\lambda^2 + \lambda + 2 & \text{for } 4 \le \lambda \le 9, \\ \frac{1}{2}(\lambda-1)(M + \sqrt{M^2 + 4(\lambda-1)}) - (\lambda - 1) & \text{for } \lambda \ge 10, \end{cases}$$

where $M = (\lambda-1)(\lambda^2 - 3\lambda + 3)$.

Not every quasi-residual design is residual. The first counterexample was given by Bhattacharya (1944). It is a 2-(16, 6, 3) design (potentially the residual of a 2-(25, 9, 3)) having two blocks which intersect in 4 points. (Of course, this cannot happen in a residual design!) For further examples, see Van Lint (1978).

After a lull, interest in quasi-symmetric designs was re-awakened by Neumaier (1982), who published a list of feasible parameter sets for small quasi-symmetric designs. His tables contained a number of question marks, whose resolution provided a challenge. This led to applications of methods of coding theory (to be discussed in Chapter 13) by Tonchev (1986) and Calderbank (1987), (1988a), (1988b). The results are fairly technical, and we refer the interested reader to the original papers.

Quasi-symmetric designs with $x = 0$ have received the most attention. We state without proof two results describing opposite extremes. The first is due to Baartmans and Shrikhande (1982), the second to Cameron (to appear).

(5.11) Theorem. *Let $\mathcal{D}$ be a quasi-symmetric 2-design with $x = 0$, having no three pairwise disjoint blocks. Then $2y \le k \le y(y+1)$; the lower bound holds if and only if $\mathcal{D}$ is a Hadamard 3-design, and the upper bound if and only if $\mathcal{D}$ is an extension of a square 2-design with the parameters given in* (1.35)(b). □

Note that the hypothesis implies that the complement of the block graph contains no triangles. We will discuss such strongly regular graphs further in Chapter 8. Baartmans and Shrikande give further numerical information too.

(5.12) Theorem. *Let $\mathcal{D}$ be a quasi-symmetric 2-design with $x = 0$, having a family of v/k pairwise disjoint blocks. Then one of the following occurs:*
(a) $\lambda = 1$;
(b) $\mathcal{D}$ is affine;
(c) $v = y(2y+1)(2y+3)$, $k = y(2y+1)$, $\lambda = y(2y-1)$. □

Of course, v/k is the maximum possible number of pairwise disjoint blocks. A resolvable quasi-symmetric design satisfies the hypothesis. Any affine design is resolvable, and many resolvable designs with $\lambda = 1$ are known, beginning with Kirkman's schoolgirls. But no example of (c) is known with $y > 1$.

By (1.43), the quasi-symmetric 4-designs are known; but the problem of determining the quasi-symmetric 3-designs is still open. It is conjectured that the only such designs with $x > 0$ and $k \leq \frac{1}{2}v$ are the 4-(23, 7, 1) design and its derived and point-residual designs. We mention a recent result of R. M. Pawale, related to (5.11), concerning the designs in the conclusions of (1.35) and (5.6)(b).

(5.13) Theorem. *A quasi-symmetric 3-design, whose block graph contains no triangles, is an extension of a square 2-design (but not a 3-(496, 40, 3) design) or the complement of one.* □

For further information, we refer to the forthcoming monograph by Sane and Shrikhande.

(5.14) DEFINITION. As we remarked before (5.6), a square design cannot be a 3-design. We say that a square design is *quasi*-3 if the number of blocks containing three points takes just two distinct values x^* and y^*, where $x^* < y^*$

Clearly these parameters satisfy $x^* \geq 0$ and $y^* \leq \lambda$. We first examine the extremal cases in these two inequalities. In the case where $y^* = \lambda$, three points lying in y^* blocks are collinear, so the Dembowski-Wagner Theorem (1.24) applies:

(5.15) Proposition. *A square quasi-3 design with $y^* = \lambda$ is a projective geometry.* □

Now let $\mathcal{D}$ be a quasi-3 design. Let $\mathcal{E} = \mathcal{D}_p^\top$ be the dual of a derived design of $\mathcal{D}$. Then $\mathcal{E}$ is a quasi-symmetric 2-$(k, \lambda, y^* - 1)$ design, in which two blocks intersect in x^* or y^* points. If $y^* < \lambda$ (as we may assume, in view of (5.15)), then $\mathcal{E}$ has no repeated blocks. This enables us to use results about quasi-symmetric designs to study quasi-3 designs.

In particular, consider the case where $x^* = 0$. Then $\mathcal{E}$ attains the bound of (5.6)(b), and so it is a 3-design. This means that three blocks of $\mathcal{D}$ intersect in 0 or y^* points, that is, the dual of $\mathcal{D}$ is also a quasi-3 design. (It is not known whether the dual of any square quasi-3 design is necessarily a quasi-3 design, though no counterexamples are known.)

In one particular case, we can obtain a characterization:

(5.16) Theorem. *A quasi-3 Hadamard 2-design is either a projective geometry over* F_2 *or the unique* 2-(11, 5, 2) *design.*

PROOF. Replace $\mathcal{D}$ by its complement, and construct the quasi-symmetric design $\mathcal{E}$ as above. This design satisfies the hypotheses of (5.8), and so it is either a Hadamard 3-design or the unique 2-(6, 3, 2) design.

Suppose that $\mathcal{E}$ is a Hadamard 3-design for any point p of $\mathcal{D}$. Translating the condition 'the complement of a block of $\mathcal{E}$ is a block' back to $\mathcal{D}$, we find that, given any two points p and q, there is a third point r such that every block containing p and q also contains r; that is, lines in $\mathcal{D}$ have size 3. But then the Dembowski-Wagner Theorem (1.24) applies. (See also Exercise 4 of Chapter 1.)

If $\mathcal{E}$ is a 2-(6, 3, 2) design, then $\mathcal{D}$ is a 2-(11, 5, 2) design. There is a unique design with these parameters (see Exercise 8(a) of Chapter 1). □

We conclude the chapter with a family of square quasi-3 designs. These also give interesting quasi-symmetric designs, by the procedure above. Our treatment follows Cameron and Seidel (1973).

(5.17) EXAMPLE. Let V be a vector space over a field F, of dimension n. We already met quadratic forms, defined in terms of coordinates; but there is an abstract definition, as follows. A function $Q : V \to \mathsf{F}$ is a *quadratic form* if the following conditions are satisfied:

(a) $Q(\alpha\mathbf{x}) = \alpha^2 Q(\mathbf{x})$ for all $\alpha \in \mathsf{F}$, $\mathbf{x} \in V$;

(b) the function $B : V \times V \to \mathsf{F}$ defined by

$$B(\mathbf{x}, \mathbf{y}) = Q(\mathbf{x} + \mathbf{y}) - Q(\mathbf{x}) - Q(\mathbf{y})$$

is bilinear.

REMARK. If $\mathsf{F} = \mathsf{F}_2$, then (a) can be replaced by the simpler statement $Q(\mathbf{0}) = 0$.

Over a field of odd characteristic (or characteristic zero), the symmetric bilinear form B determines Q, by the rule $Q(\mathbf{x}) = \frac{1}{2}B(\mathbf{x}, \mathbf{x})$. But this fails in characteristic 2. Instead, the form B is *alternating*, that is, $B(\mathbf{x}, \mathbf{x}) = 0$.

A bilinear form is *non-degenerate* if its radical is zero, that is, $B(\mathbf{x}, \mathbf{y}) = 0$ for all $\mathbf{y} \in V$ implies $\mathbf{x} = 0$. A non-degenerate alternating bilinear form can only exist on a vector space of even dimension.

We now specialize to the case where $\mathsf{F} = \mathsf{F}_2$ and the dimension n of V is even, say $n = 2m$. For a quadratic form Q, let B_Q be the alternating bilinear form obtained by polarizing Q (as in (b) of the definition). The map $Q \mapsto B_Q$ is a linear surjection from the vector space of quadratic forms to the space of alternating bilinear forms. Its kernel is the set of functions L satisfying $L(\mathbf{x}+\mathbf{y}) = L(\mathbf{x}) + L(\mathbf{y})$; that is, the linear functions on V. If

$$\mathcal{Q}_B = \{Q : B_Q = B\},$$

for an alternating bilinear form B, then $\mathcal{Q}_0$ is the dual space V^* of V, and $\mathcal{Q}_B$ is a coset of V^*.

We need the fact that, if B is non-degenerate, then $\mathcal{Q}_B$ contains $2^{2m-1} + \epsilon 2^{m-1}$ forms with $2^{2m-1} + \epsilon 2^{m-1}$ zeros in V, for $\epsilon = \pm 1$ (see Exercise 4, or (12.8), (12.9)). The number ϵ is the *type* of the quadratic form; it is essentially the same as the *Arf invariant*. We denote the type of Q by $\epsilon(Q)$.

At last we can define the designs. Choose a non-degenerate bilinear form B. The point set of the design $\mathcal{D}(m)$ is V. For each $Q \in \mathcal{Q}_B$, we associate a block

$$Y(Q) = \{\mathbf{x} \in V : (-1)^{Q(\mathbf{x})} = \epsilon(Q)\}.$$

Each block has cardinality $2^{2m-1} + 2^{m-1}$. Let $Q_1, Q_2 \in \mathcal{Q}_B$. Then $Q_1 + Q_2 = L$ is a non-zero linear form, and

$$Y(Q_1)\Delta Y(Q_2) = \{\mathbf{x} \in V : (-1)^{L(\mathbf{x})} = -\epsilon(Q_1)\epsilon(Q_2)\},$$

which has cardinality 2^{2m-1}; so the intersection of the blocks $Y(Q_1)$ and $Y(Q_2)$ has cardinality

$$\tfrac{1}{2}\left((2^{2m-1} + 2^{m-1}) + (2^{2m-1} + 2^{m-1}) - 2^{2m-1}\right) = 2^{2m-2} + 2^{m-1}.$$

It follows (for example, by (1.42)) that $\mathcal{D}(m)$ is a square 2-$(2^{2m}, 2^{2m-1}+2^{m-1}, 2^{2m-2}+2^{m-1})$ design.

Moreover, $\mathcal{D}(m)$ has the following *symmetric difference property*, due to Kantor (1975):

(Δ) The symmetric difference of any three blocks is either a block or the complement of one.

This follows because the sum of three quadratic forms in $\mathcal{Q}_B$ is another quadratic form in $\mathcal{Q}_B$. In consequence, the cardinality of the symmetric difference of three blocks takes just two values. From this, it follows that the same is true of the cardinality of the intersection of three blocks; that is, the dual of $\mathcal{D}(m)$ is a quasi-3 design.

In fact, $\mathcal{D}(m)$ is self-dual, and so is itself a quasi-3 design. This can be seen from the following alternative description.

The non-degenerate bilinear form can be used to set up an isomorphism between V and its dual space $V^* = \mathcal{Q}_0$, where the vector $\mathbf{v} \in V$ maps to the linear form $L_\mathbf{v} : \mathbf{x} \mapsto B(\mathbf{x}, \mathbf{v})$. So we can take $\mathcal{Q}_0$ as point set instead of V.

Note that $(Q+L_{\mathbf{v}})(\mathbf{x}) = Q(\mathbf{x}) + B(\mathbf{x},\mathbf{v}) = Q(\mathbf{x}+\mathbf{v}) + Q(\mathbf{v})$. Hence

$$\epsilon(Q+L_{\mathbf{v}}) = \begin{cases} \epsilon(Q) & \text{if } Q(\mathbf{v}) = 0, \\ -\epsilon(Q) & \text{if } Q(\mathbf{v}) = 1; \end{cases}$$

in other words, $\epsilon(Q+L_{\mathbf{v}}) = (-1)^{Q(\mathbf{v})}\epsilon(Q)$. So

$$\{\mathbf{v} : (-1)^{Q(\mathbf{v})} = \epsilon(Q)\} = \{\mathbf{v} : \epsilon(Q+L_{\mathbf{v}}) = +1\},$$

which translates into the set $\{L : \epsilon(Q+L) = +1\}$ under the bijection.

Thus $\mathcal{D}(m)$ can be taken to have point set $\mathcal{Q}_0$ and block set $\mathcal{Q}_B$; the rule for incidence is that $L \in \mathcal{Q}_0$ and $Q \in \mathcal{Q}_B$ are incident if and only if $\epsilon(Q+L) = 1$.

From this definition, it follows that adding a fixed form $Q \in \mathcal{Q}_B$ interchanges the sets $\mathcal{Q}_0$ and $\mathcal{Q}_B$ and induces a polarity of the design. If $\epsilon(Q) = +1$, then every point is absolute for this polarity; otherwise, no point is absolute. In the latter case, we have a polarity of the complementary design $\overline{\mathcal{D}(m)}$ with every point absolute. Thus, the graph of the appropriate null polarity (see (2.27)) is strongly regular, with parameters

$$(2^{2m}, 2^{2m-1} + \epsilon(Q)2^{m-1} - 1, 2^{2m-2} + \epsilon(Q)2^{m-1} - 2, 2^{2m-2} + \epsilon(Q)2^{m-1}).$$

We denote the graph by $\Gamma^{\epsilon}(m)$ if $\epsilon(Q) = \epsilon$.

The vertex set of these graphs can be 'translated back' to V using the inverse of the above bijection. We find that two vertices v, w of $\Gamma^{\epsilon}(m)$ are adjacent if and only if $Q(v+w) = 0$, where Q is a fixed form of type ϵ. In the case $m = 2$, these graphs are the complement of $L_2(4)$ and the Clebsch graph, respectively. In general, they have the property (which will be studied further in Chapter 8) that the induced subgraphs on both the neighbours and the non-neighbours of a vertex are strongly regular.

Using the 'translated' definition, we can also see a large group of automorphisms of the design: translation by any linear form, or any linear transformation of V preserving the form B, is an automorphism.

This definition also allows more flexibility. Instead of 0 and B, we can use any two alternating bilinear forms B_1 and B_2 whose sum is non-degenerate to determine the point and block sets. We obtain the same design, but the added flexibility will be used in Chapter 12 to construct a so-called *system of linked square designs.* We also give a re-interpretation of this construction in coding-theoretic terms.

Exercises

1. This exercise outlines a counting proof, due to C. Norman, of the fact that the block graph of a quasi-symmetric design is strongly regular.

Let B_1 and B_2 be blocks of a 2-(v, k, λ) design, whose intersection has size z. (The case $B_1 = B_2$, $z = k$, is allowed.) Let a_{ij} be the number of blocks B satisfying $|B \cap B_1| = i$, $|B \cap B_2| = j$. By counting triples p, q, B with $p \in B_1$, $q \in B_2$, and $p, q \in B$ (where $p = q$ is permitted), prove that

$$zr + (k^2 - z)\lambda = \sum_{i,j} ija_{ij}.$$

In addition, we have

$$\sum_{i,j} a_{ij} = b.$$

Now let $\mathcal{D}$ be quasi-symmetric. Show that these equations contain enough information to determine the numbers a_{ij}, first in the case $B_1 = B_2$, and then in general, in terms of z.

2. Let $\mathcal{D}$ be the 3-(22, 6, 1) design. Define a graph Γ as follows. The vertices are the pairs of points of $\mathcal{D}$. Two vertices are adjacent whenever they are disjoint but contained in a block of $\mathcal{D}$.

Prove that Γ is strongly regular, with parameters (231, 30, 9, 3).

3. Let B be a non-degenerate alternating bilinear form on a non-zero vector space V.

Prove that V has a 2-dimensional subspace U such that $V = U \oplus U^{\perp}$, and the restrictions of B to U and $U^{\perp}$ are both non-degenerate.

[Hint: Let U be spanned by $\mathbf{u}_1$ and $\mathbf{u}_2$ where $B(\mathbf{u}_1, \mathbf{u}_2) = 1$.]

Deduce that

(a) the dimension of V is even;

(b) B is unique up to invertible linear transformation of V.

Now suppose that the field is $\mathbb{F}_2$, and let $\dim(V) = 2m$. Using the above decomposition, and induction, show that a quadratic form which polarizes to B has $2^{2m-1} + \epsilon 2^{m-1}$ zeros, and that there are $2^{2m-1} + \epsilon 2^{m-1}$ such forms, where $\epsilon = \pm 1$.

4. Let H_1 and H_2 be matrices with constant row and column sums. Prove that the Kronecker product $H_1 \otimes H_2$ has constant row and column sums. (It is a Hadamard matrix: see Exercise 2 of Chapter 1.)

Recall the connection between Hadamard matrices with constant row and column sums and square 2-$(4u^2, 2u^2 + u, u^2 + u)$ designs (Exercise 3 of Chapter 1). Let H_1 be the 4×4 matrix $J - I$. Identify the design associated with the Kronecker product of m copies of H_1 with $\mathcal{D}(m)$.

[Hint: Use the decomposition of the preceding question.]

5. Show that, for $n > 2$, the points and $(n-2)$-flats of $\mathrm{PG}(n, q)$ form a quasi-symmetric design, whose block graph is isomorphic to the line graph of the point-line design of $\mathrm{PG}(n, q)$.

6. Let $\mathcal{D} = (X, \mathcal{B})$ be an affine design with m blocks in each parallel class, and $\mathcal{D}' = (X', \mathcal{B}')$ a square design with m points. Take a bijection ϕ_c from X' to each

parallel class $\mathcal{C}$ of $\mathcal{D}$, and define

$$\mathcal{D}'' = \left(X, \left\{ \bigcup_{x \in B} \phi_{\mathcal{C}}(x) : B \in \mathcal{B}', \mathcal{C} \text{ a parallel class of } \mathcal{D} \right\} \right).$$

Show that $\mathcal{D}''$ is a quasi-symmetric 2-design, and that its block graph is complete multipartite.

6. A property of the number six

In this chapter, we explore a remarkable property of the number 6. There are six objects constructed in a canonical way on a set of 6 points (in fact, 1-factorizations of the pair design). Six is the only number for which this is possible. We use this fact to give constructions and uniqueness proofs for the projective plane of order 4, the Moore graph of valency 7, and the 5-(12, 6, 1) Steiner system. This material is loosely based on lectures by G. Higman.

Let $A = \{a, b, c, d, e, f\}$ be a set of size 6. We regard A as the point set of the 2-(6, 2, 1) pair design, that is, the edge set of the complete graph K_6. We call the 2-subsets of A *edges*. A 1-factor is a set of three pairwise disjoint edges; for brevity we call this a *factor*. (Other terminology is used; for example, Sylvester (1844) called edges and factors 'duads' and 'synthemes'.)

A 1-factorization (for short, *factorization*) is a partition of the $\binom{6}{2} = 15$ edges into five factors.

(6.1) Theorem. *There are six different factorizations, any two isomorphic. Two disjoint 1-factors are contained in a unique factorization.*

PROOF. Two disjoint factors together form a union of cycles of even lengths, necessarily a single 6-cycle. A factor disjoint from these two either consists of the three long diagonals of the hexagon, or one long diagonal and the two perpendicular short ones (Fig. 6.1). Since there are three long and six short diagonals, the remaining factors in a factorization are the three of the second type. This proves the uniqueness of the factorization up to isomorphism. Now there are $15 \cdot 8 = 120$ choices of two disjoint factors; any factorization contains $5 \cdot 4 = 20$ such pairs, so there are 6 factorizations. □

Now we characterize this situation.

(6.2) Theorem. *Six is the only natural number n for which there is a construction of n isomorphic objects on an n-set A, invariant under all permutations of A, but not naturally in one-to-one correspondence with the points of A.*

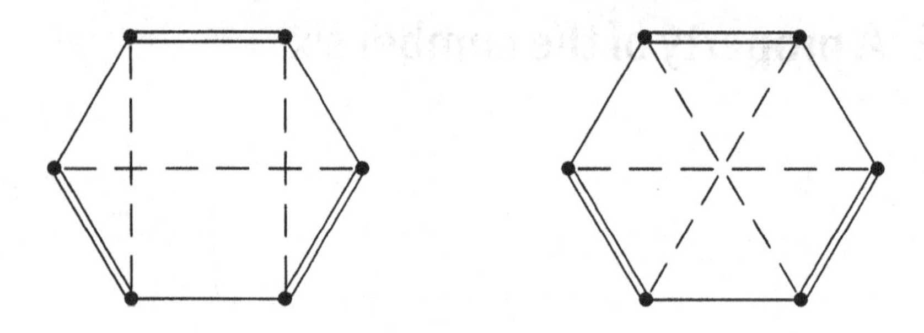

Fig. 6.1. Triples of 1-factors

PROOF. In the course of the proof, the fact that the objects are factorizations will emerge naturally.

Let G be the symmetric group on A, and let H be the automorphism group of one of the n objects. Then H is a subgroup of index n in G, which is not the stabilizer of a point of A.

Suppose first that H fixes a proper subset of A, of size k, where $0 < k < n$. (Such a group is *intransitive*.) Then, since G permutes k-sets transitively, the index of H in G is divisible by the number $\binom{n}{k}$ of k-sets. This is only possible if $k = 1$ or $n-1$ and H is a point stabilizer, contrary to assumption.

Suppose next that H fixes a partition of A into parts of size r, where r divides n, $1 < r < n$. (Such a group is *imprimitive*.) Then, as before, n is divisible by the number $n!/((r!)^n(n/r)!)$ of such partitions, which is impossible.

Suppose next that H contains a 3-cycle. For $a, b \in A$, write $a \sim b$ if either $a = b$ or $a^\tau = b$ for some 3-cycle τ. This relation is reflexive and symmetric, and is also transitive, since two 3-cycles which move a common point generate the alternating group of degree 4 or 5. This argument also shows that, if $a \sim b \sim c$, then the 3-cycle $(a\ b\ c)$ is in H. By the preceding paragraph, $\sim$ is the universal relation, and H contains all 3-cycles. But the 3-cycles generate the alternating group, which has index 2 in G; and, for $n = 2$, the alternating group fixes a point.

A similar but easier argument shows that H contains no transposition.

Now consider the $\binom{n}{2}$ transpositions in the symmetric group. These are distributed among the $n-1$ proper cosets of H. If two transpositions moving a common point, say $(a\ b)$ and $(a\ c)$, lie in the same coset of H, then $(a\ b)(a\ c)^{-1} = (a\ b\ c) \in H$, which is impossible. So each coset must contain $n/2$ pairwise disjoint transpositions defining a factor, and we obtain a factorization of A preserved by H.

Now the stabilizer of a factor has index at most $n-1$ in H, and so index at most $n(n-1)$ in G. But there are $(n-1)!! = (n-1)(n-3)\cdots 3\cdot 1$ factors, permuted transitively by G. So $n(n-1) \geq (n-1)!!$, which implies that $n \leq 6$. Since $n = 4$ is easily excluded, we are left with $n = 6$. □

This result can be expressed in various ways. In group-theoretic language, the symmetric group of degree n has an outer automorphism if and only if $n = 6$. (This is true for infinite cardinals n as well; see Schreier and Ulam (1937) for the countably infinite case.) In category-theoretic language, the category whose objects are the n-element sets and whose morphisms are the bijections between them has a non-trivial functor to itself if and only if $n = 6$. Similar ideas relate to the Axiom of Choice for families of n-element sets; see Mostowski (1945) and Exercise 9.

It follows abstractly from (6.2) that, if X denotes the set of factorizations of the 6-set A, then the set of factorizations of X is naturally in one-to-one correspondence with A. Let us see how this works. Any factor is contained in exactly two factorizations. This can be seen by observing that there are 8 ways to add a second factor to form a hexagon, and each factorization contains four of these.) Conversely, two factorizations share a unique factor. So the factors can be identified with the pairs of factorizations, that is, the edges of X. Now any edge is contained in three factors, and clearly no two of these can be contained in the same factorization; so edges correspond to factors of X. Finally, a point of A lies in five edges, no two of which are in a common factor; so points of A correspond to factorizations of X.

We call the 6-sets A and X *dual* if they stand in this relation, that is, each is bijective with the set of factorizations of the other.

We now use (6.1) to construct a projective plane of order 4. Let A be a 6-set. Define a structure whose points are the elements of A and the factors, and whose lines are the edges and the factorizations. The incidence relation holds between a point and an edge containing it, between an edge and a factor containing it, and between a factor and a factorization containing it. Now, for example, two edges either intersect in a unique point or lie together in a unique factor. A small amount of further checking shows that the structure defined is indeed a projective plane of order 4. The set A is a Type II oval in this plane.

By pushing the argument a little further, we can show the following:

(6.3) Theorem. *There is, up to isomorphism, a unique projective plane of order 4. It has the property that any four points, no three collinear, are contained in a unique oval and a unique Baer subplane.*

(The term 'oval' is used here to mean a Type II oval, that is, a set of 6 points meeting every line in 0 or 2 points. Recall that a Baer subplane is a set of 7 points meeting every line in 1 or 3 points.)

PROOF. Any projective plane contains a set of 4 points with no three collinear. Let $\{p_1, \ldots, p_4\}$ be such a set in a projective plane of order 4. Let q_1, q_2, q_3 be the intersection points of the lines p_1p_2 and p_3p_4, p_1p_3 and p_2p_4, and p_1p_4 and p_2p_3 respectively.

We claim that q_1, q_2, q_3 are collinear. Suppose not. Then p_1p_2 contains p_1, p_2, q_1 and two further points. The line q_1q_2 contains q_1, q_2, a point on p_1p_4, a point on p_2p_3, and one further point not previously mentioned. This gives us $4 + 3 + 6 \cdot 2 + 3 = 22$ distinct points, which is impossible in a 21-point structure.

Thus, $\{p_1, p_2, p_3, p_4, q_1, q_2, q_3\}$ is the unique Baer subplane containing the original four points. Also, the line $q_1q_2q_3$ contains two further points r_1 and r_2; now we have accounted for $4 + 3 + 6 \cdot 2 + 2 = 21$ points, that is, the whole plane. The only points not lying on a line with two of the p_i are r_1 and r_2; and the line r_1r_2 contains no p_i. So $\{p_1, p_2, p_3, p_4, r_1, r_2\}$ is the unique oval containing the original set.

To complete the argument, we have to show that the structure of a projective plane of order 4 containing an oval A is uniquely determined. There are $\binom{6}{2} = 15$ lines meeting A in two points; they can be labelled with the edges of A. Now a point outside A lies on three lines meeting A; the corresponding edges form a factor, which can be used as a label for the point. Finally, let L be a line disjoint from A. For any edge $\{a, b\}$, the line ab meets L in a unique point; so the labels of the 5 points of L form a factorization, which can be used to label L. Thus, we have recovered our original construction. $\square$

(6.4) REMARK. Consider the graph Γ whose vertices are the points of a projective plane of order 4 outside an oval A, two vertices adjacent if the line joining them meets A. This graph is the 15-vertex graph in (2.3)(c). (To see that condition $(*)$ of (2.3) holds, observe that the triangles are the triples of points on lines meeting A; and, if $\{x_1, x_2, x_3\}$ is a triangle and y another point, then three of the lines through y meet A, two at points of the line $x_1x_2x_3$, so y is adjacent to a unique point of the triangle.)

Next, we turn to Moore graphs. Recall from Exercises 9, 10 of Chapter 2 that a *Moore graph* of diameter 2 is a regular graph of diameter 2 and girth 5 (equivalently, a strongly regular graph with $\lambda = 2$, $\mu = 1$), and that such a graph must have valency 2, 3, 7 or 57. For $k = 2, 3$, the unique graphs are the pentagon and the Petersen graph.

Let $\{\infty, 0\}$ be an edge in the Moore graph Γ of diameter 2 and valency k, and let $A = \Gamma(\infty) \setminus \{0\}$, $X = \Gamma(0) \setminus \{\infty\}$. Any further vertex has a unique neighbour in each of the sets A and X, and so can be labelled with the pair $(a, x) \in A \times X$. Every pair in $A \times X$ labels a unique vertex. So the vertex set is $\{\infty, 0\} \cup A \cup X \cup (A \times X)$. We know all edges except for those within $A \times X$.

Furthermore, given $a_1, a_2 \in A$ and $x_1 \in X$, there is a unique common neighbour of the non-adjacent points (a_1, x_1) and a_2, that is, a unique $x_2 \in X$ such that (a_1, x_1) is joined to (a_2, x_2).

From this, we can see the uniqueness of the Petersen graph; for, if $A = \{a, b\}$ and $X = \{x, y\}$, then we must have edges $\{(a, x), (b, y)\}$ and $\{(a, y), (b, x)\}$.

We can also see, in an elementary way, the non-existence of a Moore graph of valency 4. For let $A = \{a, b, c\}$ and $X = \{x, y, z\}$. If (a, x) is joined to (b, y), then both of these vertices must be joined to (c, z), creating a triangle.

We now give the construction of the *Hoffman-Singleton graph* (1960), the unique Moore graph of valency 7. Let A and X be dual 6-sets. Note that edges of A correspond to factors of X and dually. Take the vertex set $\{\infty, 0\} \cup A \cup X \cup (A \times X)$, with the edges already specified. In addition, join (a, x) to (b, y) if and only if $\{a, b\}$ is an edge of the factor corresponding to $\{x, y\}$ (or dually). We obtain a graph of valency 7 on 50 vertices. To show that it is a Moore graph, it suffices to show that the girth is 5. We see easily that any cycle of length less than 5 must be contained in $A \times X$. The non-existence of triangles follows from our remarks on triples and partitions; the non-existence of quadrangles, by a similar argument.

Once again, we can prove uniqueness as well. This depends on the following lemma. We let $M(k)$ denote a Moore graph with diameter 2 and valency k.

(6.5) Lemma. *$M(7)$ does not contain an induced subgraph isomorphic to $M(3)$ with an edge deleted.*

We have indicated a proof of this fact in Exercise 3. □

(6.6) Theorem. *There is, up to isomorphism, a unique Moore graph of diameter 2 and valency 7.*

PROOF. Let Γ be such a graph, and (as above) write the vertex set as $\{\infty, 0\} \cup A \cup X \cup (A \times X)$, where $|A| = |X| = 6$.

Suppose that there is an edge $\{(a, x), (b, y)\}$. Then the set

$$\{\infty, 0, a, b, x, y, (a, x), (a, y), (b, x), (b, y)\}$$

contains all but one of the edges of $M(3)$. By (6.5), the remaining edge, $\{(a, y), (b, x)\}$, must also be present. We write $\{a, b\} \sim \{x, y\}$ if this occurs. Now, for any $x, y \in X$, there are three disjoint pairs of elements $a, b \in A$ for which $\{a, b\} \sim \{x, y\}$, forming a factor on A. The five factors corresponding to pairs $\{x, y\}$ with x fixed are easily seen to form a factorization. So A and X are dual sets, and we recover our previous construction. □

For our final application, we produce a 5-(12, 6, 1) design, and prove its uniqueness. First, we make one further observation. Given two disjoint factors (forming a hexagon), we observed that there is a unique factor disjoint from both but not in the factorization containing them — it consists of the three long diagonals of the hexagon. The three factors correspond to three pairs from the dual set X which pairwise intersect but do not have a common element; these must have the form $\{x, y\}$, $\{x, z\}$, and $\{y, z\}$, for some $x, y, z \in X$. Also, the complementary graph of the hexagon plus long diagonals consists of two triangles. In this way, we have a bijection between partitions of A and X into disjoint 3-sets. The dually-defined bijection is just the inverse of this one.

A permutation which acts as a transposition on A acts as a product of three disjoint transpositions on X, their cycles defining the factor corresponding to the edge on A defined by the original transposition; and dually. Thus, the combinatorial relation is reflected by the structure of the permutations. Similarly, a 3-cycle on A acts as a product of two 3-cycles on X, and dually; corresponding partitions are orbits of the same Sylow 3-subgroup of S_6.

Now we construct $S(5,6,12)$. Let A and X be dual 6-sets. Write $\{a, b\} \sim \{x, y\}$ if $\{a, b\}$ is an edge in the factor corresponding to $\{x, y\}$, and dually. Also, write $\{a, b, c\} \sim \{x, y, z\}$ if these 3-sets are parts in corresponding partitions. (Both relations are symmetric.)

Now take point set $A \cup X$, and $\mathcal{B}$ the set of blocks of the following types:
(a) A, X;
(b) $A \setminus \{a, b\} \cup \{x, y\}$, $X \setminus \{x, y\} \cup \{a, b\}$ whenever $\{a, b\} \sim \{x, y\}$;
(c) $\{a, b, c, x, y, z\}$ whenever $\{a, b, c\} \sim \{x, y, z\}$.

We claim that $(A \cup X, \mathcal{B})$ is a 5-(12, 6, 1) design. To show this, we observe first that there are

$$2 + 2 \cdot \binom{6}{2} \cdot 3 + \binom{6}{3} \cdot 2 = 132 = \binom{12}{5} \Big/ \binom{6}{5}$$

blocks; so the average number of blocks containing a 5-set is one, and it suffices to show that no two blocks meet in 5 points. If $|B_1 \cap B_2| \geq 5$, then we could assume by duality that at least three points of the intersection lie in A; then a small amount of case checking shows that this is not possible.

(6.7) Theorem. *There exists a unique* 5-(12, 6, 1) *design, up to isomorphism.*

PROOF. The intersection triangle for this design is given in Table 6.1.

We see that the complement of a block is again a block. Furthermore, the third derived design is an affine plane of order 3. So, if the point set is $A \cup B \cup C \cup D$ where $|A| = |B| = |C| = |D| = 3$ and $A \cup B$ and $A \cup C$ are blocks, then $A \cup D$ is a block. Combining these observations, $B \cup C$ is also a block. That is, if two blocks meet in 3 points, then their symmetric difference is a block.

						132						
					66		66					
				30		36		30				
			12		18		18		12			
		4		8		10		8		4		
	1		3		5		5		3		1	
1		0		2		3		2		0		1

Table 6.1.

Select a complementary pair of blocks, and call them A and X. For $a, b \in A$, the 4-set $A \setminus \{a, b\}$ lies in three blocks other than A; these have the form $A \setminus \{a, b\} \cup \{x, y\}$, where the three pairs $\{x, y\}$ form a factor of X. The complement of each of these is again a block. It is now straightforward to check that this bijection between edges of A and factors of X arises from a duality between A and X.

Now let B be a block with $B \cap A = \{a, b, c\}$, $B \cap X = \{x, y, z\}$. Then none of the three factors of A corresponding to pairs in $\{x, y, z\}$ can have an edge in $\{a, b, c\}$, or else some block would meet $B \triangle X$ in five points. So we must have the design just constructed. □

It is possible to continue the process one step further. There is a notion of a 'dual pair' of 5-(12, 6, 1) designs. This can be seen most easily from our construction of the 5-(12, 6, 1) design from the Hadamard matrix H of order 12 in Chapter 1. Let R and C be the sets of rows and columns of H. Recall that any pair of rows of H defines a partition of C into two 6-sets, and the $2 \cdot \binom{12}{2} = 132$ 6-sets arising in this way are the blocks of a 5-(12, 6, 1) design on C. Dually, pairs of columns define a design on R. We call these a *dual pair* of Steiner systems.

The duality can be seen intrinsically. Let $(R, \mathcal{B})$ be a 5-(12, 6, 1) design. We saw that the 132 blocks come in 66 disjoint pairs. there are two possible relationships between two disjoint pairs: their blocks can intersect in 3, 3, 3 and 3 points, or in 4, 2, 2 and 4. Form a graph whose vertices are the disjoint pairs of blocks, adjacency corresponding to the first relation. This graph is strongly regular, with the parameters of $T(12)$ (see Exercise 6). So, by (3.11), it is isomorphic to $T(12)$. So disjoint pairs of blocks correspond bijectively to 2-subsets from another 12-set C.

Take two points $r_1, r_2 \in R$. The blocks containing r_1 and r_2 form a 3-(10, 4, 1) design or inversive plane on the remaining 10 points. The inversive plane has the property that its 30 blocks can be split into two families of 15 so that blocks from different families intersect in an even number of points. (See Exercise 7.) So the corresponding 30 pairs of elements of C are the edges of a graph with two connected components, each having 15 edges. The components must be complete graphs on 6 vertices. Thus, a pair of elements of R corresponds to a partition of C into two 6-sets. The 132 6-sets arising in this way form the dual Steiner system on C.

Now take a 4-subset R' of R. The four blocks containing R' correspond to four disjoint edges in C, whose complement is a 4-set C'. Write $R' \sim C'$ if this holds.

Now construct a design with point set $R \cup C$. The blocks are of two types:
(a) the union of two points of R and a corresponding block of C, or dually;
(b) $R' \cup C'$, where $R' \sim C'$.
These $2 \cdot 2 \cdot \binom{12}{2} + \binom{12}{4} = 759$ 8-sets are the blocks of a 5-(24, 8, 1) design.

It is possible to show the uniqueness of the 5-(24, 8, 1) design in this way. First, show that there must exist a *dodecad*, a 12-set meeting every block in an even number (at most 6) of points. Then show that a dodecad and its complement carry a dual pair of 5-(12, 6, 1) Steiner systems, and the entire 5-(24, 8, 1) design is reconstructed in the manner given.

Unfortunately, the process does not continue; there is no dual pair of 5-(24, 8, 1) designs!

Exercises

1. Let q be an odd prime power. Let $\mathsf{F} = \mathsf{F}_q$. Define a 1-factorization of the pair design on the projective line $\{\infty\} \cup \mathsf{F}$, with 1-factors indexed by F: the factor indexed by $\alpha \in \mathsf{F}$ is

$$\{\{\infty, \alpha\}\} \cup \{\{\beta, \gamma\} : \beta + \gamma = 2\alpha\}.$$

Prove that this factorization admits the linear fractional group $\mathrm{PGL}(2, q)$ if and only if $q \leq 5$.

(Hint: $\mathrm{PGL}(2, q)$ is generated by the maps $z \mapsto cz + d$ ($c, d \in \mathsf{F}$, $c \neq 0$) and $z \mapsto 1/z$.)

In the case $q = 5$, by calculating the order of $\mathrm{PGL}(2, 5)$, show that there are six factorizations isomorphic to the given one.

2. Give an 'elementary' proof of the non-existence of a Moore graph of diameter 2 and valency 5.

3. In this exercise, we show that a Moore graph $M(7)$ of valency 7 has no induced subgraph isomorphic to $M(3)$ with an edge deleted.

Suppose that Y is a subset of the vertex set of $M(7)$ of cardinality n; let the valencies of its points (in the induced subgraph) be $r_1, \ldots, r_n$.

(1) Prove that $\sum r_i(r_i - 2) \leq n^2 - 117n + 100$.

[There are $n(n-1)$ pairs of distinct vertices in Y; of these, $\sum r_i$ are joined by edges, and $\sum r_i(r_i - 1)$ by paths of length 2 in Y; the remaining pairs have unique common neighbours outside Y. So, if the number of points outside Y with j neighbours in Y is z_j, then

$$\sum z_j = 50 - n,$$
$$\sum j z_j = 7n - \sum r_i,$$

$$\sum j(j-1)z_j = n(n-1) - \sum r_i^2.$$

Now use the fact that $\sum(j-1)(j-2)z_j$ is non-negative.]

(2) Now let Y_1 be a set of 10 vertices containing 14 edges of $M(3)$. Prove that there are five points outside Y_1 having two neighbours in Y_1. If Y consists of Y_1 together with these five points, calculate lower bounds for the valencies of the points in Y, and deduce that (1) is violated.

4. Let Γ be a Moore graph of valency k.

There are three possible relationships between two distinct edges of Γ:

(a) they meet at a vertex;
(b) not (a), but some edge meets both;
(c) neither of the above.

(Note that, if $k = 2$, then type (c) does not occur.)

Let Δ be the graph whose vertex set is the edge set of Γ, two vertices adjacent if their relation is of type (b).

Prove that the number of common neighbours of two vertices is $(k-1)^2$, $4(k-2)$, or $4(2k-5)$ according as the relation of the two points is (a), (b) or (c).

Deduce that Δ is strongly regular if and only if $k = 2, 3$ or 7.

5. Let Δ be the strongly regular graph on 175 vertices constructed as in Exercise 4 from the Hoffman-Singleton graph. Show that the parameters of Δ satisfy $k = 2\mu$. Deduce the existence of a regular two-graph on 176 vertices.

6. Let Γ be the graph whose vertices are the complementary pairs of blocks of a 5-(12, 6, 1) design, two vertices adjacent if the blocks intersect in 3 points. Prove that Γ is strongly regular, with the same parameters as $T(12)$.

7. In Chapter 2, Exercise 1, we constructed a 2-(10, 4, 2) design from the graph $L_2(3)$. (The blocks of this design are just the cliques in the regular two-graph which extends the graph — see Chapter 4.) Using the fact that $L_2(3)$ is isomorphic to its complement, find another 2-(10, 4, 2) design on the same point set, with blocks disjoint from those of the first design. Prove that

(a) the union of the two block sets forms the unique 3-(10, 4, 1) design (the inversive plane of order 3);
(b) a block of the first design meets a block of the second in 0 or 2 points.

8. Prove that there are 30 different ways of imposing the structure of a projective plane of order 2 on a given set X of seven points, falling into two orbits of 15 under the alternating group A_7.

Let U be one of these orbits of 15 planes. Prove that a 3-set is a line in exactly three planes in U, no two having any further line in common. Deduce that any two planes in U share exactly one line. Can you identify the 2-(15, 3, 1) design formed by these planes and 3-sets?

Let V be the set of 3-subsets of X. Form a graph Γ with vertex set $U \cup V$, having edges of two types, as follows:

$u \in U$ and $v \in V$ are adjacent whenever the 3-set v is a line of the plane u;

$v_1, v_2 \in V$ are adjacent whenever they are disjoint.

Prove that Γ is the Hoffman-Singleton graph.

9. If four people play all three possible doubles matches, then either one (unique) person wins all matches, or one (unique) person loses all matches. From this, it follows that, if a choice function is assumed to exist for the class of all 2-element sets in the universe, then a choice function exists for the class of all 4-element sets.

(This observation is due to Tarski: see Mostowski (1945).)

7. Partial geometries

Let s, t, α be positive integers.

(7.1) DEFINITION. A *partial geometry* with parameters s, t, α is an incidence structure (whose elements are called *points* and *lines*) having the following properties:
(a) any line is incident with $s+1$ points, and any point with $t+1$ lines;
(b) two lines are incident with at most one point (and two points with at most one line);
(c) if p is a point not incident with a line L, then there are exactly α incident point-line pairs (q, M) such that p is incident with M and L with q.

We have stated the definition in an over-elaborate way to emphasize its duality. The second, 'dual' part of (b) (in parentheses) follows from the first part. Also, (c) can be stated more briefly: if p and L are not incident, then exactly α points of L are collinear with p (or, exactly α lines on p are concurrent with L). We see that the dual of a partial geometry with parameters s, t, α is a partial geometry with parameters t, s, α.

REMARK. Clearly we have $\alpha \leq s+1$ and $\alpha \leq t+1$.

We have already met some partial geometries, intimately connected with strongly regular graphs. The first examples occurred back in (2.3), where the three graphs in part (c) of that theorem give rise to partial geometries with parameters $s = 2$, $t = u$, $\alpha = 1$, for $u = 1, 2, 4$ (we take the lines of the geometry to be the triangles in the graph). Indeed, (2.3) shows that a partial geometry with $s = 2$ and $\alpha = 1$ necessarily has $t = 1, 2$ or 4. The edges and factors of a set of six points, discussed in Chapter 6, form a partial geometry with parameters 2, 2, 1.

Partial geometries were introduced by Bose (1963) in order to provide a setting and generalization for known characterization theorems for strongly regular graphs. We will describe his results, and Neumaier's (1979) extensions of them, in this chapter. In addition, we give some properties and examples of partial geometries, and some extensions of them (in the sense of design theory).

REMARK. Bose used r, k, t for what we have called $t+1, s+1, \alpha$. Bose's r and k are the usual 1-design parameters.

(7.2) DEFINITION. The *point graph* of a partial geometry is the graph whose vertices are the points of the geometry, two vertices being adjacent whenever they are collinear. Simple counting arguments show the following result.

(7.3) Proposition. *The point graph of a partial geometry with parameters s, t, α is strongly regular, with parameters*

(7.4) $$((s+1)(st+\alpha)/\alpha,\ (t+1)s,\ s-1+t(\alpha-1),\ (t+1)\alpha).$$

□

REMARK. The lines of a partial geometry are $(s+1)$-cliques in its point graph. They are cliques of maximal size (see Exercise 1), but there may be other cliques of the same size.

Dually, the *line graph* is the graph whose vertices are the lines, adjacent if concurrent; it is just the point graph of the dual geometry, and so is also strongly regular.

A strongly regular graph is called *geometric* if it is the point graph of a partial geometry. It is *pseudo-geometric* (s, t, α) if its parameters are given by (7.4). Clearly, a geometric graph is pseudo-geometric. Bose (1963) showed that, if s is sufficiently large compared to t and α, the converse holds:

(7.5) Theorem. *Assume that*

$$s > \tfrac{1}{2}(t+2)((t-1)+\alpha(t^2+1)).$$

Then a pseudo-geometric (s, t, α) graph is geometric. □

We do not give a proof of Bose's Theorem. The method of proof is illustrated by the argument we gave for (3.11). The most difficult part is to show that any edge lies in a unique 'grand clique' of size $s+1$. Then it is not too hard to show that the points and 'grand cliques' form the required partial geometry (see Exercise 1).

We now describe some of the consequences of Bose's Theorem.

It is readily checked that a strongly regular graph with the same parameters as $T(n)$ is pseudo-geometric $(n-2, 1, 2)$. By (7.5), if $n > 8$, then such a graph is geometric. But the only partial geometry with parameters $(n-2, 1, 2)$ is the dual of the pair design on n points. In this way, we obtain Theorem (3.11). In the same way, Shrikhande's theorem characterizing $L_2(n)$ by its parameters for $n > 4$ is a consequence of (7.5).

(7.6) DEFINITION. A partial geometry with $\alpha = t$ is called a *net* of *order* $n = s+1$ and *degree* $r = t+1$.

If p is a point not on the line L of a net, then there is a unique line through p not meeting L. Just as for affine planes, it follows that the lines fall into r parallel classes of n lines each, any parallel class partitioning the n^2 points. In Chapter 1, we discussed the connection between affine planes, nets, and mutually orthogonal Latin squares. In particular, for $r \geq 2$, any r parallel classes of lines of an affine plane form a net. Such a net is called *embeddable*. Bruck (1963) proved the following result, where the *deficiency* of a net is the number $d = n + 1 - r$ by which the net falls short of being an affine plane.

(7.7) Theorem. *Suppose that*

$$n > \tfrac{1}{2}(d-1)(d^3 - d^2 + d + 2).$$

Then a net of order n and deficiency d is embeddable.

PROOF. The point graph Γ of the given net $\mathcal{N}$ is strongly regular, and so its complement is also. We find that the complement of Γ is pseudo-geometric $(n-1, d-1, d-1)$. By (7.5), if the inequality of (7.7) holds, then $\overline{\Gamma}$ is geometric, that is, it is the point graph of a net $\mathcal{N}'$ of order n and degree d. Now any pair of points lie in a unique line of either $\mathcal{N}$ or $\mathcal{N}'$, according as the points are adjacent in Γ or not. So the point set, together with the union of the line sets of $\mathcal{N}$ and $\mathcal{N}'$, is an affine plane. □

(7.8) EXAMPLE. There are two 'essentially different' Latin squares of order 4 (the addition tables of the two abelian groups of order 4). The complements of the point graphs of the corresponding nets are $L_2(4)$ (for the Klein group V_4) and the Shrikhande graph (for the cyclic group C_4). So the first, but not the second, net is embeddable.

Now we examine more closely the structure of partial geometries. Recall that $1 \leq \alpha \leq \min(s, t) + 1$.

(7.9) Proposition. *A partial geometry with $\alpha = s + 1$ is a 2-design with $\lambda = 1$.*

PROOF. Given points p and q, there is a line L containing p. If $q \notin L$ then, since $\alpha = s + 1$, every point of L (in particular, p) is collinear with q. □

REMARK. This class includes the pair design ($s = 1, \alpha = 2$), whose line graph is $T(t + 2)$.

Dually, of course, a partial geometry with $\alpha = t + 1$ is a dual 2-design.

(7.10) DEFINITION. A *complete transversal t-design* consists of a set X of kn points partitioned into k classes of size n (sometimes called 'groups', but not groups in the mathematical sense), and a set $\mathcal{B}$ of blocks, each block a transversal for the set of 'groups', with the property that any t points from different 'groups' lie in exactly λ blocks.

(7.11) Proposition. *(a) A partial geometry with $\alpha = s$ is a complete transversal 2-design with $\lambda = 1$.*

(b) A partial geometry with $\alpha = t$ is a net.

PROOF. We already saw (b); (a) is proved by dualizing, or directly. □

REMARK. As we saw in Chapter 1, nets with $t > 1$ are equivalent to sets of mutually orthogonal Latin squares. A net with $t = 1$ is simply a square grid, with point graph $L_2(n)$. Dually, the point graph of a partial geometry with $\alpha = s = 1$ is a complete bipartite graph, whose edges are the lines.

An important improvement of Bose's Theorem was proved by Neumaier (1979).

(7.12) Theorem. *Let Γ be a strongly regular graph with parameters n, k, λ, μ, whose adjacency matrix has eigenvalues k, r and s. Suppose that s is an integer less than -1, and that*

$$r \leq \tfrac{1}{2}s(s+1)(\mu+1) - 1.$$

Then Γ is the point graph of a partial geometry whose parameters satisfy $\alpha = t + 1$ or $\alpha = t$, in other words, a dual design or a net. □

Neumaier's Theorem improves on Bose's in two respects. First, it is not assumed that the graph is pseudo-geometric. (In the case of a pseudo-geometric graph, Neumaier's inequality reduces to Bose's.) This means that the theorem includes a non-existence result: strongly regular graphs whose parameters satisfy the inequality but are not pseudo-geometric do not exist. Secondly, Neumaier shows that the only partial geometries whose parameters satisfy Bose's inequality are the dual 2-designs and the nets (dual complete transversal 2-designs).

In Chapter 3, we saw that a strongly regular graph with least eigenvalue -2 is complete multipartite, or geometric, or one of finitely many exceptions. Using his result, Neumaier was able to generalize this result as follows.

(7.13) Theorem. *Let m be an integer greater than 1. Then, with finitely many exceptions, a strongly regular graph with least eigenvalue $-m$ is complete multipartite with block size m, or the line graph of a 2-design or a complete transversal 2-design with block size m.* □

In particular, all but finitely many strongly regular graphs with least eigenvalue -3 are complete multipartite with block size 3, Latin square graphs, or line graphs of Steiner triple systems. There are large numbers of non-isomorphic graphs of this type (see Wilson (1974b) for results on the number of Steiner triple systems, for example).

(7.14) DEFINITION. A partial geometry with $\alpha = 1$ is called a *generalized quadrangle* of *order* (s, t).

The parameters of a generalized quadrangle with $s > 1$ satisfy $t \leq s^2$; this is a consequence of the Krein bound (2.26), and can also be proved directly (see Exercise 3 of Chapter 8).

The commonest types of partial geometries are 2-designs, complete transversal 2-designs, and their duals. Generalized quadrangles are also fairly common, and have an extensive literature of their own (for an account of them, see Payne and Thas (1984)). We now give some examples of generalized quadrangles and of 'exceptional' partial geometries (not of any of these types).

(7.15) EXAMPLE. The 'classical' generalized quadrangles are defined as follows. Let V be a vector space over $\mathsf{F} = \mathsf{F}_q$, and B a non-degenerate form in two variables on V. We assume that B is either bilinear and *alternating* (that is, $B(\mathbf{x}, \mathbf{x}) = 0$ for all $\mathbf{x} \in V$), or *Hermitian* (that is, B is linear in its first argument, and $B(\mathbf{y}, \mathbf{x}) = B(\mathbf{x}, \mathbf{y})^\sigma$, where σ is a field automorphism of order 2 — note that the existence of σ requires q to be a square). A third situation is where V carries a non-singular *quadratic form* Q (see (5.16) for definition).

We make the further assumption that the *index* of the form B or Q (the largest dimension of a subspace of V on which the form vanishes identically) is equal to 2. Now we define an incidence structure whose points are the 1- and 2-dimensional subspaces on which the form vanishes, incidence being defined by inclusion.

It is clear that every line contains $q + 1$ points, that is, $s = q$. The constancy of t (and its value) requires further analysis, case-by-case. We show that $\alpha = 1$. Let p be a point not on the line L. If $p = [\mathbf{v}]$, consider the function $f : L \to \mathsf{F}$ defined by $f(\mathbf{x}) = B(\mathbf{x}, \mathbf{v})$. (In the case of a geometry defined by a quadratic form Q, B is obtained by polarizing Q.) Then f is linear, and is non-zero (since otherwise the form vanishes identically on the 3-dimensional space $[L, p]$). So $\ker(f)$ is a point, the unique point of L collinear with p.

The classical generalized quadrangles are called *symplectic*, *unitary* or *orthogonal* respectively in the three cases, borrowing the names given to the associated classical groups. Table 7.1 gives their parameters.

Type	dim(V)	s	t
Symplectic	4	q	q
Unitary	4	q	$q^{1/2}$
Unitary	5	q	$q^{3/2}$
Orthogonal	4	q	1
Orthogonal	5	q	q
Orthogonal	6	q	q^2

Table 7.1.

The symplectic quadrangles in dimension 4 and the orthogonal quadrangles in dimension 5 (which have the same parameters) are dual to each other in general, and are isomorphic (and self-dual) if q is a power of 2. Also, the unitary quadrangles in dimension 4 are dual to the orthogonal quadrangles in dimension 6.

(7.16) EXAMPLE. The only known generalized quadrangles with s and t greater than 1 and not both powers of the same prime, have $s = q - 1$, $t = q + 1$, where q is a prime power (or dually). The first examples were found by Ahrens and Szekeres (1969); those with q even were found independently by Hall (1971). For q even, there is a simple construction, which we present in a more general form below. See Exercise 2 for a general construction which works for both odd and even q.

Let $\mathcal{K}$ be a non-empty set of points in the projective space $\mathrm{PG}(n-1, q)$, regarded as the hyperplane at infinity of the affine space $\mathrm{AG}(n, q)$. We construct an incidence structure $T^*(\mathcal{K})$ as follows. The points of $T^*(\mathcal{K})$ are the points of $\mathrm{AG}(n, q)$; the blocks are those lines of $\mathrm{AG}(n, q)$ which meet the hyperplane at infinity in a point of $\mathcal{K}$. Clearly, there are q^n points; any line contains q points; any point lies on $|\mathcal{K}|$ lines; and any two points lie on at most one line.

Now suppose that $\mathcal{K}$ is a *set of class* $(0, m_1, m_2, \ldots)$ *with respect to lines*; in other words, any line meets $\mathcal{K}$ in 0 or m_1 or m_2 or ... points. (For brevity, we call such a set a $(0, m_1, m_2, \ldots)$-set; this terminology is not standard.) Take a line L of $T^*(\mathcal{K})$, and a point p not on L. Then $[L, p]$ is a plane, meeting the hyperplane at infinity in a line M containing at least one point of $\mathcal{K}$, viz. $L \cap \mathcal{K}$. Let p' be any other point of $M \cap \mathcal{K}$. Then the line pp' is a line of $T^*(\mathcal{K})$ meeting L (since it lies in the plane $[L, p]$). Every line meeting L arises thus. So the number of lines of $T^*(\mathcal{K})$ containing p and intersecting L is $m_1 - 1$ or $m_2 - 1$ or ....

Now, if $\mathcal{K}$ is a Type II oval (a (0,2)-set in $\mathrm{PG}(2, q)$, in the above terminology), then the argument shows that $T^*(\mathcal{K})$ is a generalized quadrangle of order $(q-1, q+1)$.

(7.17) REMARK. The line graph of a generalized quadrangle with order $(q-1, q+1)$ is a (v, k, λ) graph (in the terminology introduced in Chapter 1; that is, a strongly regular graph with $\mu = \lambda$, or the graph of a polarity of a square 2-design with no absolute points.) Moreover, it has $v = \lambda^2(\lambda + 2)$, $k = \lambda(\lambda + 1)$; so it is extremal with respect to the inequality proved in (2.28).

(7.18) EXAMPLE. More generally, a $(0, k)$-set $\mathcal{K}$ in $\mathrm{PG}(2, q)$ (with $k > 1$) gives rise to a partial geometry having parameters $s = q - 1$, $t = (q + 1)(k - 1)$, $\alpha = k - 1$. (The number of points in such a set is $1 + (q + 1)(k - 1)$, since each of the $q + 1$ lines through a point of $\mathcal{K}$ contains $k - 1$ further points of $\mathcal{K}$.)

(7.19) EXAMPLE. Another class of partial geometries is obtained from $(0, k)$-sets in projective planes. Let $\mathcal{K}$ be such a set, in a plane of order q. Define the incidence structure $T(\mathcal{K})$ whose points are the points outside $\mathcal{K}$, and whose lines are the *secants* of $\mathcal{K}$ (the lines meeting $\mathcal{K}$ in k points). Then $T(\mathcal{K})$ is a partial geometry with $s = q-k$, $t = q - (q/k)$, $\alpha = q - k - (q/k) + 1$.

In the case where $q = 4$, $k = 2$, the partial geometry is a generalized quadrangle of order $(2,2)$. We saw this embedding in Chapter 6.

A $(0,k)$-set in a projective plane is sometimes called a *maximal k-arc*. We do not use this term, since such a set is not an arc unless $k = 2$, and even in that case there are many arcs which are maximal but are not Type II ovals! A necessary condition for the existence of a $(0,k)$-set $\mathcal{K}$ in a projective plane of order q, with $k < q+1$, is that k divides q. (For we saw above that $|\mathcal{K}| = qk - q + k$; and $\mathcal{K}$ is partitioned by the secants through an exterior point, so there are $q - q/k + 1$ such secants. This calculation, incidentally, verifies the value of t in the partial geometry $T(\mathcal{K})$.)

Denniston (1969) showed that, if q is a power of 2, then $\mathrm{PG}(2,q)$ contains a $(0,k)$-set for every proper divisor k of q (see Exercise 4). On the other hand, Thas (1975) showed that, if q is a power of 3 and $q > 3$, then $\mathrm{PG}(2,q)$ contains no $(0,3)$-set (and hence no $(0,(q/3))$-set either, see Exercise 3). It is conjectured that, for odd q, $\mathrm{PG}(2,q)$ contains no $(0,k)$-set with $1 < k < q$.

(7.20) EXAMPLE. Let V be a 6-dimensional vector space over $\mathsf{F} = \mathsf{F}_3$, and Z the subspace spanned by the all-1 vector. The linear map f from V to F defined by

$$f(x_1, \ldots, x_6) = x_1 + \ldots + x_6$$

has the property that $Z \leq \ker(f)$, so it induces a map $\overline{f}$ from V/Z to F. For each $a \in \mathsf{F}$, let X_a be the set of cosets $Z + \mathrm{v}$ with $\overline{f}(Z + \mathrm{v}) = f(\mathrm{v}) = a$. Now, for $a \neq b$, we define an incidence structure with point set X_a and line set X_b, a point x and line y being incident if there are coset representatives of x and y which agree in all but one coordinate.

This a partial geometry with $s = t = 5$, $\alpha = 2$. (The values of s and t are clear. To check the condition involving α requires a small amount of case analysis. We can suppose that $a = 0$, $b = 1$, and that x is the coset containing the zero vector; so the six lines incident with x are the cosets containing the unit basis vectors. There are two types of line not incident with x, typical coset representatives being $(-1, -1, 0, 0, 0, 0)$ and $(1, 1, -1, 0, 0, 0)$; points on the first collinear with x are $(1, -1, 0, 0, 0, 0)$ and $(-1, 1, 0, 0, 0, 0)$, and on the second $(0, 1, -1, 0, 0, 0)$ and $(1, 0, -1, 0, 0, 0)$.

REMARK. We have two different partial geometries with the same point set X_0, with line sets X_1 and X_{-1} respectively, having the same point graph. Lines of the 'positive' and 'negative' partial geometries intersect in at most 2 points; they intersect in one point if and only if they are incident in the partial geometry (X_1, X_{-1}). This 'linking' resembles a similar phenomenon for square designs, to be discussed in Chapter 12.

This geometry was discovered by Van Lint and Schrijver (1981); the description given here was found by the present authors (1982).

The strongly regular graph on 175 vertices, constructed in Exercise 2 of Chapter 6 (with parameters $(175, 72, 20, 36)$) is pseudo-geometric $(4, 17, 2)$. Haemers (1981) proved that it is geometric, by finding a family of cliques with the right properties. We outline the background to the construction in Exercises 9 and 10.

One more infinite class of partial geometries will be constructed in Chapter 12.

We mention that the axioms for partial geometries have been generalized in various ways; see Feit and Higman (1964), Debroey and Thas (1978), for example.

Our final topic in this chapter concerns extensions. Any partial geometry is a 1-design; which partial geometries can be extended to 2-designs? The answer is not known in general. Thas (1986) considered extensions of known generalized quadrangles (or, more generally, those having the same orders as known examples). He proved the following result.

(7.21) Proposition. *Let q denote a prime power. Let $\mathcal{D}$ be a 2-design all of whose derived designs are generalized quadrangles of order (s,t). Then*
(a) $s \neq 1$;
(b) if $t = 1$ then $s = 2$;
(c) if $s = t = q$, then $q = 2$ or $q = 8$;
(d) if $s = q$, $t = q^2$, then $q = 2$ or $q = 8$;
(e) if $s = q^2$, $t = q$, then $q = 2$;
(f) if $s = q^2$, $t = q^3$, then $q = 2$ or $q = 4$;
(g) if $s = q^3$, $t = q^2$, then $q = 2$;
(h) if $s = q - 1$, $t = q + 1$, then q is odd;
(i) if $s = q + 1$, $t = q - 1$, then $q = 3, 5$ or 9.

PROOF. (a) If $s = 1$, then $\mathcal{D}_x$ is a complete bipartite graph for any point x. Now the sets $\{x\} \cup Y$, where Y is a block of the bipartition of $\mathcal{D}_x$, are the blocks of a 2-$(2n+1, n+1, 1)$ design, where $n = t + 1$. Such a design has $2(2n+1)/(n+1)$ blocks; but this number is never an integer for $n > 1$.

(b) Putting $n = s + 1$, we find by (1.33) that $n + 1$ divides $2n(n^2 + 1)$, or $n + 1$ divides 4. Thus $n = 3$, $s = 2$.

The other statements are similar applications of (1.33). See Exercise 5. Note that this theorem contains no assertion about existence. □

We will use the term *extended generalized quadrangle* (EGQ for short) for a 2-design all of whose derived designs are generalized quadrangles. If the quadrangles all have order (s,t), we speak of an EGQ(s,t). (WARNING: the term 'extended generalized quadrangle' is used by Cameron *et al/* (1990) in a sense which properly includes the one used here. Since this is a book about designs, and since an EGQ in the general sense which is a 2-design must be one of the structures considered here, there should be no confusion.)

(7.22) Proposition. *If an EGQ(2, t) exists, then $t = 1, 2$ or 4; there is a unique design for each value of t.*

PROOF. Let $\mathcal{D}$ be an EGQ$(2,t)$, and let B_1, B_2 be blocks of $\mathcal{D}$ meeting in two points; say $B_1 = \{a,b,c,d\}$, $B_2 = \{a,b,e,f\}$. In $\mathcal{D}_c$, there is a unique line containing e and meeting the line $\{a,b,d\}$; this line cannot contain a or b (else there is a triangle in $\mathcal{D}_a$ or $\mathcal{D}_b$). So the triple $\{c,d,e\}$ is contained in a block. Similarly, $\{c,d,f\}$ and $\{c,e,f\}$ are contained in blocks. These blocks must all coincide, or else $\{d,e,f\}$ is a triangle in $\mathcal{D}_c$. So $\{c,d,e,f\} = B_1 \triangle B_2$ is a block. Given $\mathcal{D}_a$ for one point a, this rule determines all the blocks of $\mathcal{D}$. It is also easily checked that the structure obtained in this way from any generalized quadrangle with $s = 2$ is a 2-design, and all its derived designs are generalized quadrangles. Now the result follows from (2.3). □

(7.23) EXAMPLE. Without having defined the Ahrens–Szekeres quadrangles, we will sketch Thas' proof that they are extendable! In fact, the extensions were constructed as 2-designs by Hölz (1981), and Thas (1986) recognized them as extensions of the Ahrens–Szekeres quadrangles.

Let q be an odd prime power. Let V be a 3-dimensional vector space over $\mathbb{F}_{q^2}$, and B a non-degenerate Hermitean form of index 1 on V (compare (7.15)). The point set U of the design will be the *unital* defined by B in PG$(2,q^2)$, that is, the set of 1-dimensional subspaces on which B vanishes. Then $|U| = q^3+1$. Any two points of U lie on a unique line of PG$(2,q^2)$, which contains $q+1$ points of U. These sets will be some of the blocks of the design.

Counting arguments now show that any line meets U in either 1 or $q+1$ points, and each point p of U lies on a unique 'tangent' $p^\perp$ with $p^\perp \cap U = \{p\}$.

A Baer subplane in PG$(2,q^2)$ is obtained from a coordinatization of the plane by taking only those points and lines whose coordinates lie in $\mathbb{F}_q$. Like a unital, a Baer subplane meets any line in 1 or $q+1$ points.

We call a Baer subplane Π *special* if the (unique) tangent to U at each point of $\Pi \cap U$ is a line of Π. The remaining blocks of the design are the sets $\Pi \cap U$ for all special Baer subplanes Π.

It will be convenient to have two reformulations of the concept of a special Baer subplane. First, note that the points of $\Pi \cap U$ form a conic in Π, so no three are collinear. Any four of them form a frame, which defines a coordinatization of the plane, and hence a unique Baer subplane, namely Π. Dually, Π is determined by any four of the tangent lines at points of $\Pi \cap U$. Thus, Π is invariant under the polarity $\perp$ defined by the form B. Conversely, any Baer subplane invariant under $\perp$ is special.

In terms of the form B, the characterization reads as follows: three points $[\mathbf{x}_1], [\mathbf{x}_2], [\mathbf{x}_3]$ of U lie in a special Baer subplane if and only if

$$B(\mathbf{x}_1, \mathbf{x}_2)B(\mathbf{x}_2, \mathbf{x}_3)B(\mathbf{x}_3, \mathbf{x}_1) \in \mathsf{F}_q.$$

(Note that $B(\mathbf{y}, \mathbf{x}) = B(\mathbf{x}, \mathbf{y})^q$, and $x^{q+1} \in \mathsf{F}_q$ for all $x \in \mathsf{F}_{q^2}$. It follows that the condition above is invariant under permutations of the three points, or multiplication of any generator $\mathbf{x}_i$ by a non-zero scalar.)

If the condition holds, then the special subplane Π is unique, and the remaining points of $\Pi \cap U$ are those points $[\mathbf{x}] \in U$ for which $B(\mathbf{x}, \mathbf{x}_1)B(\mathbf{x}_1, \mathbf{x}_2)B(\mathbf{x}_2, \mathbf{x})$, $B(\mathbf{x}, \mathbf{x}_2)B(\mathbf{x}_2, \mathbf{x}_3)B(\mathbf{x}_3, \mathbf{x})$ and $B(\mathbf{x}, \mathbf{x}_3)B(\mathbf{x}_3, \mathbf{x}_1)B(\mathbf{x}_1, \mathbf{x})$ all belong to F_q. In fact, if any two of these three quantities belong to F_q, then the third does also: for the product of these three quantities and the conjugate of $B(\mathbf{x}_1, \mathbf{x}_2)B(\mathbf{x}_2, \mathbf{x}_3)B(\mathbf{x}_3, \mathbf{x}_1)$ can be rearranged as the product of six quantities of the form $B(\mathbf{x}, \mathbf{y})B(\mathbf{y}, \mathbf{x})$, each of which lies on F_q by the definition of a Hermitean form. This means that, if the triples abc, abd, acd of points all lie in special Baer subplanes, then these subplanes coincide.

We show next that two points a, b of U lie in exactly $q+1$ blocks of the second kind. There are q^2-1 points on the line $a^\perp$ other than a and $(ab)^\perp$. If x is one of these, then a Baer subplane containing a, b and x and invariant under the polarity must also contain the point $(bx)^\perp$ on the line $b^\perp$. No three of $a, b, x, (bx)^\perp$ are collinear, so these four points lie in a unique Baer subplane, which is special. But a Baer subplane containing a and b contains $q-1$ points of $a^\perp \setminus \{a, (ab)^\perp\}$. So there are $(q^2-1)/(q-1) = q+1$ special Baer subplanes containing a and b.

This completes the construction of $\mathcal{D}$, and shows that it is a 2-$(q^3+1, q+1, q+2)$ design.

We claim next that, in order to prove that the derived designs of $\mathcal{D}$ are generalized quadrangles of order $(q-1, q+1)$, it suffices to show

(a) three points lie in at most one block;

(b) there do not exist distinct points a, b, c, d and blocks B, C, D with $a, b, c \in D$, $a, b, d \in C$, $a, c, d \in B$.

For (a) guarantees that, in a derived design $\mathcal{D}_a$, two points lie on at most one line, while (b) precludes triangles. For p and L in $\mathcal{D}_a$ with $p \notin L$, the average number of points of L collinear with p is

$$q^3 \cdot (q+2)(q+1)(q-1)/q^2(q+2) \cdot (q^3-q) = 1.$$

But this number can never exceed 1, so it is always 1.

Proof of (a). Two points lie on a unique line (that is, a block of the first kind), and no three points of a block of the second kind are collinear (since these blocks are conics in Baer subplanes). Moreover, our discussion of special Baer subplanes shows that three non-collinear points lie in at most one block of the second kind.

Proof of (b). Suppose that a configuration as described in (b) exists. At most one of the three blocks is of the first kind. If none is, then the comments about special Baer subplanes yield a contradiction. Suppose that a, b, c are collinear. Then the Baer subplanes containing B and C share the four points $a, d, (ad)^\perp$ and $(bc)^\perp$, with no three collinear; hence they coincide, a contradiction.

Exercises

1. (a) Use the variance trick to show that a clique C in a pseudo-geometric (s, t, α) graph has size at most $s + 1$, with equality if and only if every vertex outside C has exactly α neighbours in C.

(b) Hence show that a pseudo-geometric graph is geometric if every edge is contained in a unique clique of size $s + 1$.

2. (a) Let V be a 4-dimensional vector space over $\mathbb{F}_q$ carrying a non-degenerate alternating bilinear form B. Show that the points of $\mathrm{PG}(3, q)$ and the lines which are totally isotropic with respect to B form a GQ with $s = t = q$.

(b) Now let p be a point, and consider the geometry whose points are the points of $\mathrm{PG}(3, q)$ not perpendicular to p, with lines of two types:

the totally isotropic lines not containing p;

the non-isotropic lines containing p.

Prove that this geometry is a GQ with $s = q - 1$, $t = q + 1$.

3. Show that, if $\mathcal{K}$ is a $(0, k)$-set in a projective plane Π of order q, with $k < q+1$, then the set $\mathcal{K}^\top$ of lines disjoint from $\mathcal{K}$ is a $(0, (q/k))$-set in the dual plane $\Pi^\top$. Moreover, the partial geometry $T(\mathcal{K}^\top)$ (see Example (7.19)) is the dual of $T(\mathcal{K})$.

4. This exercise outlines Denniston's construction of $(0, k)$-sets in $\mathrm{PG}(2, q)$, where q is even and k divides q.

(a) Let A be the additive group of $\mathbb{F}_q$, where q is a power of 2. Show that the map $x \mapsto ax^2 + bx$ $(a, b \neq 0)$ is a homomorphism of A with kernel $\{0, b/a\}$; so the image of this map is a subgroup A_0 of A with index 2 in A. Deduce that the image of the map $x \mapsto ax^2 + bx + c$ is a coset of A_0, each value taken twice.

(b) Now let $Q(x, y)$ be an irreducible quadratic form over $\mathbb{F}_q$, and K a subgroup of A of order k, where k divides q. Prove that the set

$$S = \{(x, y) \, : \, Q(x, y) \in K\}$$

meets every line of $\mathrm{AG}(2, q)$ in 0 or k points. [The restriction of Q to a line through the origin is one-to-one. For a line not containing the origin, we have the situation in (a); either $A_0 \supseteq K$, in which case the proper coset consisting of values of Q is disjoint from K, or else $|A_0 \cap K| = |(A \setminus A_0) \cap K| = \frac{1}{2}k$, and each value is taken twice.]

(c) Add a line at infinity to obtain a $(0, k)$-set in $\mathrm{PG}(2, q)$.

5. (a) Let $\mathcal{D}$ be an $\mathrm{EGQ}(s, t)$. If the point p is not in the block B, show that B is partitioned by the blocks through p which meet it in two points. Deduce that s is even.

(b) Using this fact, complete the proof of (7.20).

6. Let U be a unital in $\mathrm{PG}(2, q^2)$, with q odd, as in Example (7.22). Let $\mathcal{B}$ be the set of triples $\{[\mathbf{x}_1], [\mathbf{x}_2], [\mathbf{x}_3]\}$ of points of U for which $B(\mathbf{x}_1, \mathbf{x}_2)B(\mathbf{x}_2, \mathbf{x}_3)B(\mathbf{x}_3, \mathbf{x}_1)$ is a square in $\mathbb{F}_{q^2}$. Prove that

(a) the defining condition for $\mathcal{B}$ is invariant under permutations of the three points, or replacement of a generator $\mathbf{x}_i$ with a scalar multiple $\alpha\mathbf{x}_i$;

(b) $(X, \mathcal{B})$ is a regular two-graph.

(These examples are due to Taylor (1977).)

7. Let $\mathcal{G} = (X, \mathcal{B})$ be the (unique) generalized quadrangle of order (4, 2). Then $\mathcal{G}$ is a 1-(45, 5, 3) design. The following construction, due to Cameron (1991), produces 2-designs $\mathcal{D}$ with $\mathcal{D}_x = \mathcal{G}$ for *one* point x.

Construct two types of auxiliary structure as follows:

- *L-structures.* One of these is a partition of the ten 2-subsets of a 5-set into five pairs, each pair consisting of two disjoint 2-sets. Show that
  - (a) each point of the 5-set is not covered by exactly one of the five pairs;
  - (b) there are exactly six different ways of imposing an L-structure on a set of five points. (These can be obtained from the six factorizations of the pair design on six points discussed in Chapter 6, by deleting a single point.)
- *P-structures.* One of these is the dual of an affine plane of order 2. The points are partitioned into three sets of size 2 (corresponding to the parallel classes of lines in the affine plane), and each block is a transversal to these three classes. Prove that, given a set of six points which is already partitioned into three 2-sets, there are exactly two ways of imposing a P-structure with the given partition.

Now take the geometry $\mathcal{G}$. For each line L of $\mathcal{G}$, impose an L-structure $\mathcal{L}(L)$ on L. Now consider a point p of $\mathcal{G}$. There are three lines L_1, L_2, L_3 containing p. In $\mathcal{L}(L_i)$, there is a unique pair of 2-sets not covering p; let P_{i1}, P_{i2} be these pairs. Now consider the 6-set $\{P_{ij} : i = 1, 2, 3;\ j = 1, 2\})$, partitioned into three 2-sets by the lines L_i $(i = 1, 2, 3)$. Impose a P-structure $\mathcal{P}(p)$ on this set so as to induce the given partition.

Now, let ∞ be a point not in X, and let $\mathcal{E}$ be the incidence structure with point set $X \cup \{\infty\}$, having blocks of two types:

(a) all sets $L \cup \{\infty\}$ for $L \in \mathcal{B}$;

(b) all unions of three 2-sets forming a block in some $\mathcal{P}(p)$.

Prove that $\mathcal{E}$ is a 2-(46, 6, 3) design and that $\mathcal{E}_\infty = \mathcal{G}$.

Show further that the $6^{27}.2^{45}$ ways of assigning the structures give rise to at least

$$6^{27} \cdot 2^{45}/46 \cdot |\mathrm{Aut}(\mathcal{G})| > 10^{28}$$

non-isomorphic designs.

Finally, show that a similar construction produces an 2-design extending any generalized quadrangle of order $(2t, t)$, provided that the auxiliary structures exist. (Such quadrangles are known only for $t = 1, 2, 4$. For $t = 1$, there is a unique extension, which we have met in several guises. You may like to estimate the number of non-isomorphic 2-(298, 10, 5) designs which can be produced by this construction.)

8. A strongly regular graph with parameters (28, 15, 6, 10) is pseudo-geometric (3, 4, 2). By (4.15), we know all the strongly regular graphs with these parameters; they are the complements of $T(8)$ and the three Chang graphs. Which of these graphs are geometric? (See De Clerck (1979).)

9. Define a distance function on the edge set E of the Petersen graph by setting

$$d(e_1, e_2) = \begin{cases} 0 & \text{if } e_1 = e_2; \\ 1 & \text{if } e_1 \text{ and } e_2 \text{ have a common vertex;} \\ 2 & \text{if an edge meets both } e_1 \text{ and } e_2; \\ 3 & \text{if none of the above hold.} \end{cases}$$

(This is just the usual distance in the line graph of the Petersen graph.)

Show that the five edges of a 1-factor are mutually at distance 2. Hence show that there are exactly six 1-factors, any edge in two, and any two edges at distance two in exactly one. Deduce that two 1-factors share a common edge.

Does the Petersen graph have a 1-factorization?

10. Suppose that $\mathcal{P}$ is a collection of Petersen subgraphs of the Hoffman–Singleton graph Γ with the property that any pentagon is contained in a unique member of $\mathcal{P}$. (Such a set would contain one-fifth of all the Petersen subgraphs of Γ.) Prove that the incidence structure whose points are the edges of Γ, and whose lines are the 1-factors in all members of $\mathcal{P}$, is a partial geometry $(4, 17, 2)$, whose point graph is the graph of Chapter 6, Exercise 2 (with $k = 7$).

REMARK. A set $\mathcal{P}$ with the required properties was found by Haemers (1981).

8. Graphs with no triangles

Strongly regular graphs with $\lambda = 0$, $\mu = 1$ are Moore graphs of diameter 2. As we have seen, there are only finitely many of them. What about graphs satisfying one of these conditions? Retaining the condition $\mu = 1$ leads to the so-called *Moore geometries*, studied by Kantor (1977) and others. In this chapter, we follow the other direction, and consider graphs with $\lambda = 0$, $\mu > 1$. These graphs are intimately connected with certain kinds of 2-designs. One of the principal results is an inequality for the parameters of such a graph, with the property that, if the bound is attained, then the 2-design is a 3-design. We conclude the chapter by pointing out that there is a generalization of this inequality to all strongly regular graphs; it is just the Krein bound, which we met in Chapter 2. Equality in the Krein bound has a similar interpretation in general. This leads to a natural strengthening of the condition of strong regularity.

As in Chapter 5, we will have to deal with 2-designs whose parameters k and λ have nothing to do with those of the graphs; so, temporarily, we use (n, a, c, d) for the parameters of a strongly regular graph. (Since these graphs will have no triangles, we will have $c = 0$.)

Let Γ be a strongly regular graph with parameters $(n, a, 0, d)$. We have $d = a$ if and only if Γ is complete bipartite; this case is trivial and will not be considered further. Also, if $d = 1$ then Γ is a Moore graph, and its valency is $a = 2, 3, 7$ or possibly 57. We regard this case as being fairly well understood, and ignore it as well. Thus, we assume that $1 < d < a$.

Let x be any vertex of Γ. Let $\mathcal{D}(\Gamma, x)$ denote the structure with point set $\Gamma(x)$ and block set $\overline{\Gamma}(x)$, incidence being defined by adjacency in Γ.

(8.1) Proposition. *$\mathcal{D}(\Gamma, x)$ is a 2-structure with parameters $v = a$, $k = d$, $\lambda = d - 1$, $r = a - 1$, $b = a(a-1)/d$, possibly with repeated blocks. Two vertices of $\overline{\Gamma}(x)$ which are adjacent correspond to disjoint blocks.*

PROOF. All this is immediate from the definition and the fact that there are no triangles; for example, any two vertices of $\Gamma(x)$ are non-adjacent, and so have d common neighbours, of which one is x and the remainder lie in $\overline{\Gamma}(x)$. The existence of two adjacent blocks with a common point would also create a triangle. □

In view of this, we will now use the design parameters v and k in place of a and d. Note that $\lambda = k - 1$.

Let us apply the integrality conditions. Type I can only occur if the graph is the pentagon, which we have excluded from consideration. Assuming Type II, we find that $k^2 - 4k + 4v$ is a perfect square. Since it is greater than k^2 and of the same parity, we can put $k^2 - 4k + 4v = (k + 2s)^2$ for some positive integer s. Then we obtain:

(8.2) Proposition. *$v = (s+1)k + s^2$, and*

$$\frac{1}{2}\left(\frac{((s+1)k+s^2)((s+2)k+s^2-1)}{k} \pm \frac{((s+1)k+s^2)((s+2)k+s^2-3)}{k+2s}\right)$$

are non-negative integers. □

In particular, the two terms inside the brackets are both integers. From this, we conclude that k divides $s^2(s^2-1)$ and $k+2s$ divides $s(s+1)(s+2)(s+3)$. The second condition implies that $k+2s$ divides $k(k-2)(k-4)(k-6)$; so, if $k \neq 2, 4, 6$, then s is bounded by a function of k, and there are only finitely many graphs with given k. (This observation is due to Biggs (1971).) For $k = 2, 4$ or 6, however, the integrality condition permits infinitely many values of s. For example, if $k = 2$, then we have $v = (s+1)^2 + 1$, and $s+1$ is not divisible by 4.

Having chosen a potential 2-$(v, k, k-1)$ design $\mathcal{D} = (X, \mathcal{B})$ for $\mathcal{D}(\Gamma, x)$, we can begin constructing the graph: the vertex set is $\{\infty\} \cup X \cup \mathcal{B}$, where ∞ is a new object; ∞ is joined to every vertex in X; a vertex in X and one in $\mathcal{B}$ are joined precisely when they are incident. Only the edges within $\mathcal{B}$ remain to be determined; and we have the restrictions that only disjoint pairs of blocks can be adjacent, and each block must be adjacent to $v - k$ others.

Only four graphs satisfying our conditions are known. We now examine each of these graphs from our present perspective.

(8.3) EXAMPLE. The Clebsch graph. Here $\mathcal{D}$ is the 2-(5, 2, 1) pair design on 5 points. Since only three pairs are disjoint from a given one, the structure of the graph is determined!

(8.4) EXAMPLE. The Gewirtz graph. $\mathcal{D}$ is the pair design on 10 points. These 10 points can be regarded as forming the projective line over $\mathbb{F}_9$, and the joining rule for disjoint pair can be expressed in terms of cross ratio, as follows: $\{a, b\}$ and $\{c, d\}$ are joined whenever

$$\frac{(a-c)(b-d)}{(a-d)(b-c)}$$

is a primitive fourth root of unity in $\mathbf{F}_9$.

(8.5) EXAMPLE. The complement of the block graph of the 3-(22, 6, 1) design (see Table 5.1). $\mathcal{D}$ is a 2-(16, 4, 3) design, defined as follows. Let V be a 4-dimensional vector space carrying a non-degenerate alternating bilinear form B. The points of the design are the vectors of V; the blocks are all cosets of 2-dimensional subspaces of V on which the form B is identically zero. Two blocks are adjacent if they are disjoint but not parallel. (Two blocks are parallel of they are cosets of the same subspace.)

(8.6) EXAMPLE. The *Higman–Sims graph* (Higman & Sims (1968)). In this case, $\mathcal{D}$ is the unique 3-(22, 6, 1) design (which, by (1.4), is also a 2-(22, 6, 5) design). According to Table 1.1, there are exactly 16 blocks disjoint from any block. Since $\overline{\Gamma}(x)$ has valency $22 - 6 = 16$, the entire graph is again determined! We will examine this remarkable situation further (and prove that the Higman–Sims graph is strongly regular) below.

In (1.20), we proved a lower bound for the number of blocks not disjoint from a given block in a 2-design. Using this, we obtain the following result.

(8.7) Theorem. *If a strongly regular graph with parameters $(n, v, 0, k)$ exists, then*

$$v \geq \tfrac{1}{2}\left(3k + 1 + (k-1)\sqrt{4k+1}\right).$$

If $v = (s+1)k + s^2$ (as in (8.2)), then $k \leq s(s+1)$.

PROOF. The number of blocks of $\mathcal{D}(\Gamma, x)$ is $v(v-1)/k$. So, by (1.20),

$$\frac{v(v-1)}{k} \geq 1 + \frac{k(v-2)^2}{k^2 - 3k + v} + (v-k).$$

So

$$k(k^2 - 3k + v) \leq (v-k)(v-k-1),$$

which gives the result. □

This result excludes such values of (v, k) as (9, 4) and (256, 48), which satisfy the integrality condition.

We now turn to the case of equality in (8.7). We saw in Chapter 5 that, of the three statements about a 2-design $\mathcal{D}$ with $2 < k < v - 1$,

(a) $\mathcal{D}$ is a 3-design,
(b) $\mathcal{D}$ is quasi-symmetric with $x = 0$,
(c) $b = v(v-1)/k$,

any two together imply the third. For our designs $\mathcal{D}(\Gamma, x)$, (c) holds, and so (a) and (b) are equivalent.

(8.8) Theorem. *Let Γ be strongly regular with parameters $(n, v, 0, k)$, where $2 < k < v$. Then the following are equivalent:*
(a) $\mathcal{D}(\Gamma, x)$ is a 3-design;
(b) $\mathcal{D}(\Gamma, x)$ is quasi-symmetric;
(c) $\Gamma|\overline{\Gamma}(x)$ is strongly regular;
(d) $v = \frac{1}{2}\left(3k + 1 + (k-1)\sqrt{4k+1}\right)$.

(8.9) REMARK. The Gewirtz graph satisfies (b) but none of the other conditions; so the hypothesis that $k > 2$ is essential.

PROOF. The remarks before the theorem show that (a) and (b) are equivalent. If either holds, then $\mathcal{D}(\Gamma, x)$ is an extension of a square 2-design, and (1.35) gives the following possibilities for the parameters:
(a) $v = 4y$, $k = 2y$;
(b) $v = y(y^2 + 3y + 1)$, $k = y(y + 1)$;
(c) $v = 496$, $k = 40$, $y = 4$.
Of these, (a) is excluded by (8.7), and (c) fails the rationality condition (8.2). (The possibility $v = 112$, $k = 12$, $y = 2$ also fails the rationality condition; the computation of Lam *et al.* (1983) is not required.) In case (b), the equality (d) of the theorem is easily verified. So (a) implies (d).

If (d) holds, then the bound in (8.6) is attained. We conclude that
(a) $\mathcal{D}(\Gamma, x)$ is quasi-symmetric (by (1.20));
(b) vertices in $\Gamma(x)$ are adjacent if and only if (as blocks) they are disjoint.
So $\Gamma|\overline{\Gamma}(x)$ is the complement of the block graph of $\mathcal{D}(\Gamma, x)$, and hence strongly regular, by (5.3). So (d) implies all the other conditions in the Theorem.

Finally, suppose that (c) holds. Let $\Gamma|\overline{\Gamma}(x)$ have parameters $(b, v-k, 0, \mu)$. Then two non-adjacent blocks are adjacent to k further vertices of Γ, of which μ are blocks; so they intersect in $k - \mu$ points. Adjacent blocks are disjoint, because there are no triangles. So $\mathcal{D}(\Gamma, x)$ is quasi-symmetric. Thus (c) implies (b). □

There is a partial converse, generalizing the Higman–Sims construction.

(8.10) Theorem. *Let $\mathcal{D} = (X, \mathcal{B})$ be a 3-$(y(y^2 + 3y + 1), y(y + 1), y - 1)$ design, for some $y > 1$ (that is, an extension of a square 2-design in case (b) of (1.35)). Let Γ be the graph with vertex set $\{\infty\} \cup X \cup \mathcal{B}$, in which ∞ is joined to all vertices in X, a vertex in X and a vertex in $\mathcal{B}$ are joined if they are incident, and two vertices in $\mathcal{B}$ are joined if they are disjoint. Then Γ is strongly regular.*

PROOF. Straightforward counting. □

This construction has a curious consequence concerning extendablity of square 2-designs. We observed in Chapter 1 that any Hadamard 2-design is uniquely extendable. What about the second case, the 2-$((\lambda+2)(\lambda^2+4\lambda+2), \lambda^2+3\lambda+1, \lambda)$ designs?

For $\lambda = 1$, there is a unique such design, the projective plane of order 4, and it has three different (though isomorphic) extensions, corresponding to its three classes of ovals. For $\lambda = 2$, there are four known 2-(56, 11, 2) designs. (The first has a polarity whose associated graph is the Gewirtz graph (see also Hall *et al.* (1970)); the others, one self-dual and one dual pair, are due to Assmus, Mezzaroba and Salwach (1977) and Denniston (1979).) However Bagchi (1988, 1991) proved that no 2-(56, 11, 2) design can be extendable. Concerning larger values of λ, we have the following result.

(8.11) Corollary. *If a 2-$((\lambda+2)(\lambda^2+4\lambda+2), \lambda^2+3\lambda+1, \lambda)$ design is extendable, then so is its dual.*

PROOF. Let $\mathcal{D}$ be the extension, with added point p. Perform the construction of (8.10) to obtain a strongly regular graph Γ. By (8.8), $\mathcal{D}(\Gamma, p)$ is a 3-design; and it is easily seen to be an extension of the dual of $\mathcal{D}_p$. □

Within a strongly regular graph with the parameters of (8.8), further graphs and designs can be found. If p and q are adjacent vertices, the, $\overline{\Gamma}(p)$ and $\overline{\Gamma}(p) \cap \overline{\Gamma}(q)$ carry strongly regular graphs with parameters $((y^2+3y+1)(y^2+2y-1), y^2(y+2), 0, y^2))$ and $(y(y+2)(y^2+2y-1), y(y^2+y-1), 0, y(y-1))$ respectively (the complements of the block graphs of $\mathcal{D}$ and its point residual $\mathcal{D}^q$ respectively). In this way, we find the 77- and 56-point graphs as subgraphs of the 100-point Higman–Sims graph. Also, if p and r are non-adjacent, then $(\Gamma(p) \setminus \Gamma(r), \Gamma(r) \setminus \Gamma(p))$ (with incidence defined by adjacency) is a square 2-$(y^2(y+2), y(y+1), y)$ design. (This is the design $\mathcal{E}^0$ in the proof of (1.35).)

We summarize in Table 8.1 the parameters of the known strongly regular graphs with no triangles, including Moore graphs but excluding complete bipartite graphs. For ease of reference, we revert to the convential notation for the parameters n, k, λ, μ; also, k, r, s are the eigenvalues. The 77-graph is the complement of the block graph of the 3-(22, 6, 1) design.

Graph	n	k	λ	μ	r	s
Pentagon	5	2	0	1	$\frac{1}{2}(\sqrt{5}-1)$	$\frac{1}{2}(-\sqrt{5}-1)$
Petersen	10	3	0	1	1	-2
Clebsch	16	5	0	2	1	-3
Hoffman–Singleton	50	7	0	1	2	-3
Gewirtz	56	10	0	2	2	-4
77-graph	77	16	0	4	2	-6
Higman–Sims	100	22	0	6	2	-8

Table 8.1.

We turn now to a wide generalization of (8.7), to arbitrary strongly regular graphs. The starting point of this generalization is the following observation, whose proof involves nothing but algebraic manipulation:

(8.12) Proposition. *For strongly regular graphs with no triangles, the inequality (8.7) is equivalent to the Krein bound (2.26)(b).* □

Of course, the Krein bound has an algebraic proof; the elementary counting argument yielding (8.7) is not available in general. Most of the results in the remainder of this chapter have algebraic proofs, which we will not give.

The *subconstituents* of a strongly regular graph Γ are defined to be the regular graphs $\Gamma|\Gamma(x)$ and $\Gamma|\overline{\Gamma}(x)$, for all vertices x. By (8.8), equality in (8.7) is characterized by the fact that the subconstituents $\Gamma|\overline{\Gamma}(x)$ are strongly regular (and, of course, the other subconstituents are null). If we temporarily extend the definition of strongly regular graphs to include complete and null graphs, we could ask: Which strongly regular graphs have all subconstituents strongly regular; in particular, is this condition equivalent to equality in the Krein bound? (Note that both of these hypotheses are preserved if the graph is replaced by its complement.)

(8.13) Example. The question in the second form given above has a negative answer. The graphs $L_2(n)$, for $n > 2$, have strongly regular subconstituents, but do not meet the Krein bound. The same is true of the graphs $\Gamma^\epsilon(m)$ of (5.17), for $m \geq 2$ (except for the Clebsch graph, which does attain the bound).

This question was considered by Cameron, Goethals and Seidel (1978a). To state their results, we need some further definitions.

A *pseudo-Latin square graph* is a pseudo-geometric graph (s, t, α) with $\alpha = t$ (see Chapter 7). In other words, it is a strongly regular graph whose eigenvalues k, r, s satisfy $k = -s(r - s - 1)$. If we put $r - s = m$ and $s = -d$, then the parameters are $(m^2, d(m-1), d^2 - 3d + m, d(d-1))$. Such a graph is denoted by $PL_d(m)$. If it is geometric, then it is the point graph of a net of order m and degree d, and is called a *Latin square graph*, denoted $L_d(m)$. (Thus, the Shrikhande graph is a $PL_2(4)$ graph.)

Mesner (1965) observed that, if we replace m and d by their negatives in the formulae giving the parameters for $PL_d(m)$, we obtain a set of parameters satisfying the Integrality Condition, viz. $(m^2, d(m+1), d^2 + 3d - m, d(d+1))$. (Note that we must have $m \leq d^2 + 3d$, since λ is non-negative.) A graph with these parameters is said to be of *negative Latin square type*, and denoted $NL_d(m)$.

A *conference graph* is one satisfying Case I of the Integrality Condition, that is, with parameters $(4\mu + 1, 2\mu, \mu - 1, \mu)$. Note that, if the number of vertices is a square, then a conference graph is simultaneously of pseudo- and negative Latin square types; conversely, a graph which is of both of these types is a conference graph.

(8.14) Proposition. *A strongly regular graph, whose valency is equal to the multiplicity of a non-principal eigenvalue, is of pseudo- or negative Latin square type or a conference graph.* □

A *Smith graph* is a strongly regular graph with parameters

$$n = \frac{2(r-s)^2((2r+1)(r-s)-3r(r+1))}{(r-s)^2-r^2(r+1)^2},$$
$$k = \frac{-s((2r+1)(r-s)-r(r+1))}{(r-s)+r(r+1)},$$
$$\lambda = \frac{-r(s+1)((r-s)-r(r+3))}{(r-s)+r(r+1)},$$
$$\mu = \frac{-(r+1)s((r-s)-r(r+1))}{(r-s)+r(r+1)}.$$

or the complement of one. (The name refers to the work of Margaret Smith (1975), who considered rank 3 graphs with these parameters.) Note that the non-negativity of λ and μ shows that necessarily $-s \geq r(r+2)$.

Finally, an *imprimitive graph* is either a disjoint union of complete graphs, or the complement of one (a complete multipartite graph). Any other strongly regular graph is called *primitive.*

Now, the main results of Cameron, Goethals and Seidel are summarized in the following two theorems.

(8.15) Theorem. *(a) A primitive strongly regular graph attains the Krein bound if and only if it is the pentagon or a Smith graph.*

(b) The subconstituents of a Smith graph are strongly regular. □

(8.16) Theorem. *Let Γ be a strongly regular graph, and suppose that its subconstituents $\Gamma|\Gamma(x)$ and $\Gamma|\overline{\Gamma}(x)$ are both strongly regular, for some vertex x. Then Γ is the pentagon, a pseudo- or negative Latin square graph, or a Smith graph.* □

(8.17) REMARK. The graphs in Example (8.13) are pseudo- or negative Latin square graphs. But not all such graphs have strongly regular subconstituents (see Exercise 7). The problem of finding the ones which do is open.

As we said, we do not give proofs of these theorems. (To get an idea of the flavour of the proofs, see Exercise 6 of Chapter 4.) But one feature of the proof is worth noting. There is a concept of 'spherical t-design' for finite sets of points on the Euclidean unit sphere, developed by Delsarte, Goethals and Seidel (1977). Let S be a subset of the unit sphere, with the property that the pairs of points in S make only two angular distances α and β. We can form a graph Γ by joining pairs of points at distance α. then it can be shown that S is a spherical 2-design if and only if Γ is strongly regular, and S is a spherical 3-design if and only if Γ attains the Krein bound. This gives a nice analogue of (8.8)! See also Cameron, Goethals and Seidel (1978a, b).

We now give two applications of these results. The first concerns generalized quadrangles.

For pseudo-geometric $(s, t, 1)$ with $s > 1$, the Krein bound is equivalent to the assertion that $t \leq s^2$ (compare Exercise 3). Now Cameron, Goethals and Seidel (1978a) showed the following:

(8.18) Proposition. *A pseudo-geometric $(s, s^2, 1)$ graph is geometric.*

PROOF. Since the Krein bound is attained, (8.15) shows that the subconstituents are strongly regular. The first subconstituent $\Gamma|\Gamma(x)$ has $s(s^2+1)$ vertices and has valency $s-1$. So, if it is strongly regular, then (2.5) shows that it has '$\mu = 0$', and so it is a disjoint union of s^2+1 complete graphs of size s. Thus, every edge of Γ lies in a clique of size $s+1$, and Γ is geometric. □

It is notable that, apart from the point graphs of generalized quadrangles with $t = s^2$, there are (up to complementation) only four known Smith graphs: the Clebsch graph, the Higman–Sims graph, the McLaughlin graph, and the 162-vertex subconstituent of the McLaughlin graph. In their paper, Cameron, Goethals and Seidel prove that each of these, as well as the pseudo-geometric $(s, s^2, 1)$ graphs for $s = 2, 3$ (the Schläfli graph and the other subconstituent of McLaughlin's graph) are characterized by their parameters. Table 8.2 gives these parameters.

Graph	n	k	λ	μ	r	s
Clebsch	16	5	0	2	1	−3
Schläfli	27	10	1	5	1	−5
Higman–Sims	100	22	0	6	2	−8
1st subconst. of McL	112	30	2	10	2	−10
2nd subconst. of McL	162	56	10	24	2	−16
McLaughlin	275	112	30	56	2	−28

Table 8.2.

The second application concerns 'higher regularity conditions' for graphs, which have a design-like flavour. Let t be a positive integer. The graph Γ is said to be ***t*-tuple regular** if, for any set S of vertices with $|S| \leq t$, the number of common neighbours of S depends only on the isomorphism type of the induced subgraph on S. It follows immediately from the definition that a graph is 1-tuple regular if and only if it is regular, and is 2-tuple regular if and only if it is strongly regular (possibly complete or null). Moreover, an inclusion-exclusion argument shows that the complement of a t-tuple regular graph is also t-tuple regular (Exercise 11).

(8.19) Proposition. *A graph is 3-tuple regular if and only if it is strongly regular with strongly regular subconstituents.*

PROOF. The constancy of λ and μ for the subconstituent $\Gamma|\Gamma(x)$ follows from the definition of 3-tuple regularity in the cases where S is a triangle or a path of length 2 respectively. The same argument applied to $\overline{\Gamma}$ handles the other subconstituent.□

Thus, (8.16) describes 3-tuple regular graphs.

The absolute bound (2.23) for strongly regular graphs shows that the eigenvalues of a Smith graph satisfy $-s \leq r^2(2r+3)$. We define an *extremal Smith graph* to be a Smith graph in which $-s = r^2(2r+3)$, or the complement of such a graph. More generally, the following result holds:

(8.20) Proposition. *A primitive strongly regular graph attains the absolute bound if and only if it is the pentagon or an extremal Smith graph.* □

Now the following theorem is due to Buczak (1980) and Cameron (1980a):

(8.21) Theorem. *(a) A graph is 4-tuple regular if and only if it is a regular disjoint union of complete graphs, a regular complete multipartite graph, the pentagon, $L_2(3)$, or an extremal Smith graph.*

(b) A graph is 5-tuple regular if and only if it is a regular disjoint union of complete graphs, a regular complete multipartite graph, the pentagon, or $L_2(3)$. All these graphs are t-tuple regular for every t.

The proof of this theorem involves solving a jigsaw puzzle. By (8.16), the graph and its subconstituents are imprimitive, pseudo- or negative Latin square type, the pentagon, or Smith graphs; it is necessary to check how the parameters can be fitted together. A simpler puzzle of the same kind is given as Exercise 12 below. □

Exercises

1. Let $(X, \mathcal{B})$ be the 3-(22, 6, 1) design, and Γ the Higman–Sims graph with vertex set $\{\infty\} \cup X \cup \mathcal{B}$. Let X_0 be a 7-subset of X forming a block in a 1-point extension (that is, meeting every block in 1 or 3 points), and $\mathcal{B}_0$ the set of blocks meeting X_0 in one point.

(a) Prove that the induced subgraph of Γ on $\{\infty\} \cup X_0 \cup \mathcal{B}_0$ is the Hoffman–Singleton graph.

(b) Show that any induced Hoffman–Singleton subgraph of Γ containing ∞ arises in this way.

2. Let Γ be the Higman–Sims graph, and Γ_0 an induced Hoffman–Singleton subgraph, with vertex sets V and V_0 respectively; let $V_1 = V \setminus V_0$.

(a) Use the variance trick to prove that the induced subgraph Γ_1 on V_1 has valency 7. [Let x_i be the number of vertices in V_1 with i neighbours in $V_0 \ldots$]

(b) Prove that Γ_1 has girth 5. [Choose $p \in V_1$, and let y_i be the number of vertices $q \in V_1$, not adjacent to p, such that p and q have i common neighbours in V_0

...]

(c) Deduce that Γ_1 is also isomorphic to the Hoffman–Singleton graph.

3. Let Γ be a strongly regular graph containing no induced subgraph isomorphic to 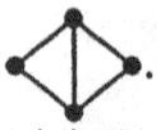.

(a) Prove that $\Gamma|\Gamma(x)$ is a disjoint union of complete graphs of size $\lambda + 1$.

(b) Define a structure $\mathcal{D}^*(\Gamma x)$, whose points are the cliques in $\Gamma(x)$ and whose blocks are the vertices in $\overline{\Gamma}(x)$, incidence being defined by adjacency. Show that it is a 2-structure.

(c) Derive an inequality similar to (8.7) by the same technique. [Hint: if p and q are non-adjacent vertices, let x_i be the number of vertices r, non-adjacent to p and q, for which p, q and r have exactly i common neighbours; apply the variance trick.]

(d) Show that the point graph of a generalized quadrangle satisfies the hypothesis, and deduce that a generalized quadrangle of order (s,t) satisfies $t \leq s^2$ if $s > 1$.

4. Let x be a vertex of the strongly regular graph Γ with parameters (n, a, c, d), where $c > 0$ and $d > 0$. Let $\mathcal{D}(\Gamma, x)$ have point set $\Gamma(x)$ and block set $\overline{\Gamma}(x)$, with incidence defined by adjacency. Prove that $\mathcal{D}(\Gamma, x)$ is a 2-structure if and only if $\Gamma|\Gamma(x)$ is strongly regular with parameters $(a, c, c - d + \frac{c(d-1)}{a-1}, \frac{c(d-1)}{a-1})$.

5. The preceding exercise gives the possibility of a strongly regular graph with parameters (26, 10, 3, 4), in which $\Gamma|\Gamma(x)$ is the Petersen graph, and $\mathcal{D}(\Gamma, x)$ is a 2-(10, 4, 2) design.

Let Γ_0 be the Petersen graph, with vertex set X. Let $\mathcal{B}$ be the set of 4-subsets of X for which the induced subgraph is the disjoint union of two edges. Prove that $\mathcal{D} = (X, \mathcal{B})$ is a quasi-symmetric 2-(10, 4, 2) design. (We have seen this design several times before in different guises; do you recognize it?)

It remains to find a joining rule for blocks, so that each block is joined to six others. The block graph of $\mathcal{D}$ has valency 6. Prove that this graph does *not* work. Can you find one that does?

(Paulus (1973) found six different examples of strongly regular graphs with parameters (26, 10, 3, 4) having this property.)

6. Prove that the condition $m \leq d(d+3)$ for $NL_d(m)$ graphs is equivalent to the Krein bound.

7. Prove that, if a pseudo-Latin square graph $PL_3(m)$ has strongly regular subconstituents, then $m \leq 4$.

8. Show that the graph $\Gamma^\epsilon(m)$ of (5.17) is of pseudo-Latin square type $PL_{t+1}(2t)$ if $\epsilon = +1$, and of negative Latin square type $NL_{t-1}(2t)$ if $\epsilon = -1$, where $t = 2^{m-1}$.

9. The strongly regular graph on 231 vertices constructed in Exercise 2 of Chapter 5 is pseudo-geometric (10, 2, 1). Prove that it is not geometric.

10. Find an infinite family of 3-tuple but not 4-tuple regular graphs whose subconstituents are all 3-tuple regular.

11. Use the Principle of Inclusion and Exclusion (1.57) to show that, if a graph is t-tuple regular, then so is its complement.

12. (a) If Γ is t-tuple regular (with $t > 1$), then its subconstituents are $(t-1)$-tuple regular.

(b) Hence prove, by induction, that a graph which is t-tuple regular for *all* t is a disjoint union of complete graphs, a complete multipartite graph, the pentagon, or $L_2(3)$.

(This result is due to Gardiner (1976), though it is stated there with stronger hypotheses, viz. that any isomorphism between induced subgraphs of Γ can be extended to an automorphism of Γ. This group-theoretic condition bears the same relation to the hypothesis of (8.16)(b) or this exercise as 'rank 3 graph' does to 'strongly regular graph'. This is a rare case where global symmetry is forced by a purely combinatorial regularity assumption.)

9. Codes

In this chapter we introduce *Coding Theory.* This topic, also known as the theory of *error-correcting codes*, has its origin in communication theory. Applications are concerned with several situations in which 'coded' messages are transmitted over a so-called *noisy* channel that has the effect that symbols in 'words' of the message are sometimes changed to other symbols of the 'alphabet'. The system is designed in such a way that the most likely error-patterns (at the receiver end) can be recognized and corrected. In this book these practical applications are of no concern. During the development of the discipline of coding theory it turned out that several results from design theory could be used to construct 'good' codes. Later, theorems from coding theory contributed considerably to design theory. These connections are what interests us here and therefore the subject will be introduced as an (abstract) area of mathematics.

In coding theory one considers a set F of q distinct symbols which is called the *alphabet.* In practice q is generally 2 and $\mathsf{F} = \mathsf{F}_2$. In most of the theory one takes $q = p^r$ (p prime) and $\mathsf{F} = \mathsf{F}_q$. The code is called a q-ary code (binary for $q = 2$, ternary for $q = 3$).

Using the symbols of F, one forms all n-tuples, that is, F^n, and calls these n-tuples *words* and n the *word length.* If $\mathsf{F} = \mathsf{F}_q$, we shall denote the set of all words by F_q^n and interpret this as n-dimensional vector space over the field F. Sometimes we omit the index and speak of the space F^n.

In F^n we introduce a distance function d (called *Hamming distance*) which is the natural distance function to use when one is interested in the number of errors in a word that is spelt incorrectly.

(9.1) DEFINITION. For $\mathbf{x} \in \mathsf{F}^n$ and $\mathbf{y} \in \mathsf{F}^n$ we denote by $d(\mathbf{x}, \mathbf{y})$ the number of coordinate places in which $\mathbf{x}$ and $\mathbf{y}$ differ, that is,

$$d(\mathbf{x}, \mathbf{y}) = |\{i : 1 \le i \le n; \quad x_i \ne y_i\}|.$$

The following two definitions are directly connected with d and the language of metric spaces.

(9.2) DEFINITION. For $\mathbf{x} \in \mathbf{F}^n$ we define the *weight* $w(\mathbf{x})$ of $\mathbf{x}$ by $w(\mathbf{x}) := d(\mathbf{x}, \mathbf{0})$. (As usual $\mathbf{0}$ denotes the zero vector in $\mathbf{F}^n$.)

(9.3) DEFINITION. For $\rho > 0$ and $\mathbf{x} \in \mathbf{F}^n$ we define the *ball* of radius ρ with centre at $\mathbf{x}$ by

$$B(\mathbf{x}, \rho) := \{\mathbf{y} \in \mathbf{F}^n : d(\mathbf{x}, \mathbf{y}) \le \rho\}.$$

REMARK. Balls are commonly referred to as 'spheres' in the literature; for example, the result of Exercise 1 is the *sphere-packing bound.* We have preferred the mathematicians' usage.

Consider a subset C of $\mathbf{F}^n$ with the property that any two distinct words of C have distance at least $2e + 1$. If we take any $\mathbf{x}$ in C and change t coordinates, where $t \le e$ (that is, we make t errors), then the resulting word still resembles the original word more than it resembles any of the others (that is, it has a smaller distance to $\mathbf{x}$ than to other words of C). Therefore, if we know C, we can correct the t errors. In coding theory such a set C is called an e-error-correcting code. Formally we define:

(9.4) DEFINITION. An *e-error-correcting code* C is a subset of $\mathbf{F}^n$ with the property

$$\forall_{\mathbf{x} \in C} \forall_{\mathbf{y} \in C} \left[\mathbf{x} \ne \mathbf{y} \Rightarrow d(\mathbf{x}, \mathbf{y}) \ge 2e + 1\right].$$

This definition implies that balls of radius e around distinct codewords are disjoint. If these balls also have the property that they cover the space (a rare but very interesting property), then the code is called *perfect.*

(9.5) DEFINITION. An e-error-correcting code C in $\mathbf{F}^n$ is called *perfect* if

$$\bigcup_{\mathbf{x} \in C} B(\mathbf{x}, e) = \mathbf{F}^n.$$

Many of the codes that we shall study have the property that as a subset of the space of words they are linear:

(9.6) DEFINITION. A k-dimensional linear subspace C of $\mathbf{F}_q^n$ is called a *linear code* or $[n, k]$ *code* over the field $\mathbf{F}_q$.

If it is not known whether C is linear, but we do know that $|C| = M$, then we use the notation (n, M) code. If d is known we write $[n, k, d]$ code or (n, M, d) code.

The error-correcting capacity of a code is determined by the *minimum distance* between all pairs of distinct codewords.

(9.7) Proposition. *The minimum distance of a linear code is equal to the minimum weight among all nonzero codewords.*

PROOF. If $\mathbf{x} \in C$, $\mathbf{y} \in C$, then $\mathbf{x} - \mathbf{y} \in C$ and $d(\mathbf{x}, \mathbf{y}) = d(\mathbf{x} - \mathbf{y}, \mathbf{0}) = w(\mathbf{x} - \mathbf{y})$. □

We shall now look at two ways of describing a linear code C. The first is given by a *generator matrix* G, which is a matrix for which the rows are a set of basis vectors of the linear subspace C. This means that $C = \{\mathbf{a}G : \mathbf{a} \in \mathsf{F}^k\}$. (Again k is the dimension of the code.) We shall call two codes *equivalent* if one is obtained from the other by applying a fixed permutation to the positions for all codewords (e.g. reversing the order of the coordinates in all codewords). One of the properties that interest us most, namely the minimum distance of a code, does not change under such a permutation. It is now clear that for any linear code there is an equivalent code that has a generator matrix G of the form

$$G = (I_k \quad P), \tag{9.8}$$

where I_k is the k by k identity matrix and P is a k by $n - k$ matrix. This is called the *standard form* for the generator matrix.

(9.9) REMARK. Sometimes the definition of equivalence is extended by also allowing a fixed permutation acting on the alphabet, or even possibly different permutations acting on the alphabet symbols in each coordinate position. (This most general group of equivalences is referred to as the *wreath product* of the permutation groups on alphabet symbols and on positions.) Such transformations preserve the cardinality and minimum distance of the code; but, in general, linearity may be lost. However, if the permutations of the alphabet which are permitted are restricted to be multiplications by non-zero field elements, then a code is transformed into an equivalent code by acting on it with a monomial matrix (a matrix with exactly one non-zero entry in each row or column) instead of a permutation matrix, and so a code equivalent (in this sense) to a linear code is linear. This form of equivalence, which we shall refer to as 'monomial equivalence', is sometimes important (see Exercise 12).

A code C is called *systematic* if there is a k-subset of the coordinate places such that to each possible k-tuple of entries in these places there corresponds exactly one codeword. By (9.8) a linear code is systematic.

We now come to the second description of a linear code C.

(9.10) DEFINITION. If C is a linear code of dimension k, then

$$C^\perp := \{\mathbf{x} \in \mathsf{F}^n : \forall_{\mathbf{y} \in C}[\langle \mathbf{x}, \mathbf{y} \rangle = 0]\},$$

where $\langle \mathbf{x}, \mathbf{y} \rangle$ denotes the usual inner product or dot product in F^n, is called the *dual code* of C.

The code $C^\perp$ is an $[n, n-k]$ code. If H is a generator matrix for the dual code $C^\perp$, then H is called a *parity check matrix* for C. In general the term 'parity check' is used for any word $\mathbf{x} \neq \mathbf{0}$ that is orthogonal to all the words of the code. (The term comes from considering the binary case; a word belongs to C if and only if it has an even number of 1s in those positions where the 1s occur in any parity check for C.) We shall call any matrix H for which the rows generate $C^\perp$ a parity check matrix (that is, we will allow H to have dependent rows in some cases). Such a parity check matrix H then defines the code C by

$$C = \{\mathbf{x} \in \mathbb{F}^n : \mathbf{x}H^\top = \mathbf{0}\}. \tag{9.11}$$

The reader is warned that often two codes C and C^* are called duals when actually C^* is equivalent to the dual of C. Although this has some advantages it also causes confusion!

A code is called *self-dual* if $C = C^\perp$. If $C \subset C^\perp$, then the code C is called *self-orthogonal.*

Let us consider an important example of linear codes. Consider the lines through the origin in $AG(m, q)$. Along each of these lines choose a vector $\mathbf{x}_i$ $(i = 1, 2, \ldots, n := (q^m - 1)/(q-1))$. So we have a representation of the points of $PG(m-1, q)$. Form the matrix H (m rows, n columns) with the vectors $\mathbf{x}_i$ as column vectors. Since no two columns of H are linearly dependent, we see that H is a parity check matrix for a code C in which every nonzero word $\mathbf{x}$ has weight at least 3. In fact the minimum weight of C is clearly 3. The code C has dimension $n - m$. The code is called the $[n, n-m, 3]$ *Hamming code* over $\mathbb{F}_q$.

(9.12) Theorem. *Hamming codes are perfect codes.*

PROOF. Let C be a $[n, n-m]$ Hamming code over $\mathbb{F}_q$. Then the balls of radius 1 around codewords are disjoint. Each such ball contains $1 + n(q-1) = q^m$ words of $\mathbb{F}_q^n$. The number of codewords is q^{n-m}. So, together the balls of radius 1 around codewords cover the whole space. □

Let C be any code. To every word $(c_1, c_2, \ldots, c_n)$ of C we adjoin an extra letter c_0 (say in front) such that $c_0 + c_1 + \ldots + c_n = 0$. In earlier terminology we could say that we require the all-one vector $\mathbf{1}$ to be a parity check. The symbol c_0 is called an *overall parity check (symbol).* In this way we obtain a code $\overline{C}$ which is called the *extended code* corresponding to C. If C is linear and H is a parity check matrix for C, then

$$\overline{H} := \begin{pmatrix} 1 & 1 & 1 & \ldots & 1 & 1 \\ 0 & & & & & \\ 0 & & & & & \\ \vdots & & & H & & \\ 0 & & & & & \\ 0 & & & & & \end{pmatrix}$$

is a parity check matrix for $\overline{C}$. Of course if C already has $\mathbf{1}$ as a parity check vector, then the extension is trivial because $c_0 = 0$ for all words of C. However, if C is a binary code with odd minimum distance d, then obviously the minimum distance of $\overline{C}$ is $d+1$.

For our first connection to the theory of designs we consider as an example the extended [8,4] binary Hamming code. We have seen above that

$$\overline{H} := \begin{pmatrix} 1 & 1 & 1 & 1 & 1 & 1 & 1 & 1 \\ 0 & 0 & 0 & 0 & 1 & 1 & 1 & 1 \\ 0 & 0 & 1 & 1 & 0 & 0 & 1 & 1 \\ 0 & 1 & 0 & 1 & 0 & 1 & 0 & 1 \end{pmatrix}$$

is a parity check matrix for this code. It is easily seen that the code is equivalent to a code with generator $G = (I_4\, J_4 - I_4)$. If we make a list of the sixteen words of this code, we find $\mathbf{0}$, $\mathbf{1}$, and fourteen words of weight 4. Now, since two words of weight 4 have distance at least 4, they have at most two 1s in common. It follows that no word of weight 3 is a 'subword' of more than one codeword. There are $\binom{8}{3} = 56$ words of weight 3 and each codeword of weight 4 has four subwords of weight 3. So all words of weight 3 are covered once.

(9.13) Proposition. *The fourteen words of weight* 4 *in the extended* [8, 4] *Hamming code form a* 3-(8, 4, 1) *design.* □

Besides as an example, the discussion above has served a second purpose, namely to show that it can be useful to know how many words of some fixed weight i are contained in a code C. A simple way to describe such information is given in the following definition.

(9.14) DEFINITION. Let C be a code of length n and let A_i $(i = 0, 1, \dots, n)$ denote the number of codewords of weight i. Then

$$A(\xi, \eta) := \sum_{i=0}^{n} A_i \xi^i \eta^{n-i}$$

is called the *weight enumerator* of C.

The weight enumerators of a code C and its dual $C^\perp$ are related. The relation is due to F. J. MacWilliams (1963).

(9.15) Theorem. *Let $A(\xi, \eta)$ be the weight enumerator of an $[n, k]$ code over $\mathbb{F}_q$ and let $A^\perp(\xi, \eta)$ be the weight enumerator of the dual code. Then*

$$A^\perp(\xi, \eta) = q^{-k} A(\eta - \xi, \eta + (q-1)\xi).$$

PROOF. Let χ be any nontrivial character of $(\mathbb{F}_q, +)$ (that is, a homomorphism from the additive group of $\mathbb{F}_q$ to the multiplicative group of non-zero complex numbers). We will use the easily verified property that

$$\sum_{c \in \mathbb{F}_q} \chi(c) = 0.$$

We denote $\mathbb{F}_q^n$ by $\mathcal{R}$. We define $f(\mathbf{v}) := \xi^{w(\mathbf{v})}\eta^{n-w(\mathbf{v})}$. Furthermore, we define

$$g(\mathbf{u}) := \sum_{\mathbf{v} \in \mathcal{R}} \chi(\langle \mathbf{u}, \mathbf{v} \rangle) f(\mathbf{v}).$$

Then we have

$$\sum_{\mathbf{u} \in C} g(\mathbf{u}) = \sum_{\mathbf{v} \in \mathcal{R}} f(\mathbf{v}) \sum_{\mathbf{u} \in C} \chi(\langle \mathbf{u}, \mathbf{v} \rangle).$$

Here the inner sum is $|C|$ if $\mathbf{v} \in C^\perp$. On the other hand, if $\mathbf{v} \notin C^\perp$, then as $\mathbf{u}$ runs through C the inner product in the inner sum takes on every value in $\mathbb{F}_q$ the same number of times. So then the inner sum is 0. This proves that

$$\sum_{\mathbf{u} \in C} g(\mathbf{u}) = |C| \cdot A^\perp(\xi, \eta).$$

For $a \in \mathbb{F}_q$ we write $w(a) := 1$ if $a \neq 0$ and $w(0) := 0$. Then we have

$$g(\mathbf{u}) = \sum_{v_1 \in \mathbb{F}_q} \cdots \sum_{v_n \in \mathbb{F}_q} \xi^{w(v_1)+\ldots+w(v_n)} \eta^{(1-w(v_1))+\ldots+(1-w(v_n))} \chi(u_1 v_1 + \ldots + u_n v_n)$$

$$= \prod_{i=1}^{n} \left\{ \sum_{v \in \mathbb{F}_q} \xi^{w(v)} \eta^{1-w(v)} \chi(u_i v) \right\}.$$

Since the inner sum is $\eta + (q-1)\xi$ if $u_i = 0$ and it is

$$\eta + \xi \sum_{a \in \mathbb{F}_q^*} \chi(a) = \eta - \xi$$

if $u_i \neq 0$, we find

$$g(\mathbf{u}) = (\eta - \xi)^{w(\mathbf{u})} (\eta + (q-1)\xi)^{n-w(\mathbf{u})}.$$

Therefore we are done. □

In Chapter 13 we shall show how this equation played an essential rôle in the proof of the nonexistence of the projective plane of order 10.

For the sake of completeness we mention that generalizations of (9.14) to nonlinear codes are known (cf. MacWilliams, Sloane, and Goethals (1972) and Delsarte (1973)).

We give one example of a nonlinear code constructed by using design theory. It is not difficult to show (see Exercise 6) that a binary code of length 10 with minimum distance 5 can have at most twelve codewords. To construct a code meeting this bound we take the rows of the incidence matrix of the 2-(11,5,2) Paley design and adjoin the all-one word **1** to this set. These twelve binary words then clearly have mutual distance 6. By deleting a coordinate we obtain a code with the required properties.

The reader interested in nonlinear codes is referred to MacWilliams and Sloane (1977).

Exercises

1. Prove that if a binary $(n, M, 2e+1)$ code exists, then

$$M \le 2^n \Big/ \left(\sum_{i=0}^{e} \binom{n}{i} \right).$$

2. Show that the parameter set (6,9,3) satisfies the conditions of Exercise 1 but that a binary code with these parameters does not exist.

3. Let C be an $[n, k]$ code over $\mathbb{F}_q$ that is systematic on any k coordinate positions. Show that $d = n - k + 1$. Such a code is called a *maximum distance separable* code (MDS code). Prove that if C is MDS, then $C^\perp$ is also MDS.

4. Let C be an $[n, k, d]$ binary code. Consider a generator matrix G of the form

$$\begin{pmatrix} 1 \;\; 1 \;\; \dots \;\; 1 \;\; 1 & 0 \;\; 0 \;\; \dots \;\; 0 \;\; 0 \\ A & B \end{pmatrix},$$

where the first row has weight $w < 2d$, A has size $k-1$ by w, and B has size $k-1$ by $n-w$. The matrix B generates a binary code C'. Show that C' is a $[n-w, k-1, d']$ code with $d' \ge d - \frac{1}{2}w$. If $w = d$ the code C' is called a *residual* of C.

Show that a binary $[n, n-7, 5]$ code does not exist if $n > 12$. (Note that the condition of Exercise 1 is satisfied for $n \le 15$.)

It is also possible but more tricky to show that a [12,5,5] code does not exist (but in Chapter 11 we shall construct a (12,32,5) binary code). To do this one shows that B is essentially determined and that it is not possible to find a matrix A satisfying the requirements.

5. Let the matrix A be the M by n matrix with all the codewords of an (n, M, d) code C over $\mathbb{F}_q$ as its rows. Estimate in two ways the sum of the distances of all pairs of codewords (or rows). Use this to show that

$$M \le \frac{d}{d - n(1 - q^{-1})}.$$

6. Show that a binary $(10, M, 5)$ code has $M \le 12$.

7. Let C be the dual of an $[n, n-3]$ Hamming code over $\mathbb{F}_p$, where $n = p^2 + p + 1$. Show that any two words of C have distance p^2. Compare this result with Exercise 5. Determine the weight enumerator of the corresponding Hamming code.

8. Let C be a binary perfect e-error-correcting code of length n with $\mathbf{0} \in C$. Consider the words of weight $2e+1$ in C as characteristic functions of subsets of the set $\{1, 2, \ldots, n\}$. Prove that this is a Steiner system $(e+1)$-$(n, 2e+1, 1)$.

9. Let C be a perfect e-error-correcting code over $\mathbb{F}_q$ with $\mathbf{0} \in C$. Prove that the weight enumerator of C is uniquely determined.

10. Show that an [8,4] self-dual code over $\mathbb{F}_p$ exists.

11. Use Theorem (9.14) to prove that a ternary self-dual code of length 10 does not exist.

12. This exercise generalizes the construction of the Hamming codes in (9.12), and also introduces an important correspondence principle.

Let $S = \{p_1, \ldots, p_n\}$ be a set of n distinct points in $\mathrm{PG}(k-1, q)$, which lie in no proper subspace. Take $\mathbf{v}_1, \ldots, \mathbf{v}_n$ to be vectors spanning the subspaces $p_1, \ldots, p_n$; let H be the matrix with columns $\mathbf{v}_1, \ldots, \mathbf{v}_n$; and let C be the code with parity check matrix H.

(a) Show that C is an $[n, n-k, \geq 3]$ code.

(b) Show that multiplying H on the left by an invertible matrix does not change C, and has the effect of applying an element of the group $\mathrm{PGL}(k, q)$ of linear collineations to S.

(c) Show that multiplying H on the right by a monomial matrix replaces C by a 'monomial equivalent' linear code (as defined in Remark (9.9)), and does not affect S.

REMARK. Thus, we have a bijection between 1-error-correcting linear codes (up to monomial equivalence), and spanning sets of points in projective space (up to linear collineations). The Hamming codes arise in the case when S consists of all the points of $\mathrm{PG}(k-1, q)$.

(d) Show that supports of words of weight 3 in C correspond to collinear triples in S, while zero sets of words in $C^\perp$ correspond to hyperplane sections of S.

(e) Choose a familiar configuration of points in projective space. What properties does the corresponding code have?

10. Cyclic codes

Many of the most interesting codes that we shall study are *cyclic*. We define these as follows.

(10.1) DEFINITION. An $[n, k]$ code C over $\mathbb{F}_q$ is called *cyclic* if

$$\forall_{(c_0,c_1,\ldots,c_{n-1})\in C}[(c_{n-1}, c_0, c_1, \ldots, c_{n-2}) \in C].$$

From now on we make the restriction $(n, q) = 1$. (Much of the theory goes through without this restriction, but one gains very little by dropping it.) Let $R := \mathbb{F}_q[x]$ be the ring of all polynomials with coefficients in $\mathbb{F}_q$ and let S be the ideal generated by $(x^n - 1)$. Clearly the residue class ring R/S (considered as an additive group) is isomorphic to $\mathbb{F}_q^n$. An isomorphism is given by

$$(a_0, a_1, \ldots, a_{n-1}) \leftrightarrow a_0 + a_1x + \ldots + a_{n-1}x^{n-1}$$

since it is obvious that the polynomials of degree less than n form a set of representatives for R/S. From on we do not distinguish between *words* and *polynomials* of degree $< n \pmod{(x^n - 1)}$. Note that multiplication by x in R/S amounts to the cyclic shift $(a_0, a_1, \ldots, a_{n-1}) \rightarrow (a_{n-1}, a_0, a_1, \ldots, a_{n-2})$.

From this it follows that a cyclic code C corresponds to an *ideal* (which we also denote by C) in R/S. Since this ring is a principal ideal ring, every cyclic code in $\mathbb{F}_q^n$ is a principal ideal generated by some polynomial $g(x)$ that divides $x^n - 1$. We shall call $g(x)$ the *generator (-polynomial)* of the cyclic code.

Let $g(x) = g_0 + g_1x + \ldots + g_{n-k}x^{n-k}$ be the generator of a cyclic code C and let $h(x) := (x^n - 1)/g(x) = h_0 + h_1x + \ldots h_kx^k$. The words $g(x)$, $xg(x), \ldots, x^{k-1}g(x)$ form a basis of the code C, that is,

$$G := \begin{pmatrix} g_0 & g_1 & \cdots & g_{n-k} & 0 & 0 & \cdots & 0 \\ 0 & g_0 & g_1 & \cdots & g_{n-k} & 0 & \cdots & 0 \\ \vdots & \ddots & \ddots & & \ddots & \ddots & & \vdots \\ 0 & 0 & \cdots & 0 & g_0 & g_1 & \cdots & g_{n-k} \end{pmatrix} \qquad (10.2)$$

is a generator matrix for C. Since $g(x)h(x) = 0$ in the ring R/S, we see that

$$H := \begin{pmatrix} 0 & 0 & \cdots & 0 & h_k & \cdots & & h_1 & h_0 \\ 0 & & \cdots & h_k & & \cdots & h_1 & h_0 & 0 \\ \vdots & & & & & & & & \vdots \\ h_k & & \cdots & h_1 & h_0 & 0 & \cdots & 0 & 0 \end{pmatrix}$$

is a parity check matrix for C. This gives a translation into the terminology of Chapter 9. We call $h(x)$ the *check polynomial* of the cyclic code C. Note that the *dimension* of the cyclic code is equal to the degree of the check polynomial. The code generated by $h(x)$ is obtained from the dual of C by reversing the order of the symbols. It is often considered as the dual code (see the remark following 9.10).

Very often cyclic codes are described, not by giving the generator $g(x)$, but by *prescribing certain zeros* of all codewords (in a suitable extension field of $\mathbb{F}_q$). Clearly this is an equivalent description because the requirement that all codewords are multiples of $g(x)$ means that they are 0 in the zeros of $g(x)$. To show how simple this description can be we give an example.

Let $q = 2$, $n = 2^m - 1$ and let α be a *primitive element* of $\mathbb{F}_{2^m}$. Let $m_1(x) = (x - \alpha)(x - \alpha^2)\ldots(x - \alpha^{2^{m-1}})$ be the minimal polynomial of α. We wish to consider the cyclic code generated by $m_1(x)$. Every element of $\mathbb{F}_{2^m}$ can be expressed uniquely as $\sum_{i=0}^{m-1} \epsilon_i\alpha^i$ where $\epsilon_i \in \mathbb{F}_2$. The nonzero elements of $\mathbb{F}_{2^m}$ are the powers α^j $(j = 0, 1, \ldots, 2^m - 2)$. Construct a matrix H for which the j-th column is $(\epsilon_0, \epsilon_1, \ldots, \epsilon_{m-1})^\top$ if $\alpha^j = \sum_{i=0}^{m-1} \epsilon_i\alpha^i$ $(j = 0, 1, \ldots, 2^m - 2)$. If $\mathbf{a} := (a_0, a_1, \ldots, a_{n-1})$, $a(x) := a_0 + a_1 + \ldots a_{n-1}x^{n-1}$, then the vector $\mathbf{a}H^\top$ corresponds to the field element $a(\alpha)$. Therefore $\mathbf{a}H^\top = \mathbf{0}$ means the same thing as $m_1(x)|a(x)$. Since the columns of H are a permutation of the binary representations of $1, 2, \ldots, 2^m - 1$, we have proved the following theorem.

(10.4) Theorem. *The binary cyclic code of length $n = 2^m - 1$ for which the generator is the minimal polynomial of a primitive element of $\mathbb{F}_{2^m}$ is equivalent to the $[n, n - m]$ binary Hamming code.*

EXAMPLE. Take $q = 2$, $n = 7$, $g(x) = 1 + x + x^3$. In this case, the cyclic code C generated by $g(x)$ has dimension 4. It is equivalent to the [7,4] Hamming code which corresponds to our example following (9.11). We remark that the seven cyclic shifts of the first row of (10.2) for this example form the incidence matrix of PG(2, 2). The code consists of $\mathbf{0}$, $\mathbf{1}$, these seven words, and their complements. This code is linked to an interesting question linking design theory with coding theory. Suppose a Steiner triple system $\mathcal{S}$ (like PG(2, 2) in our example) is given. Is there a linear code C over some field $\mathbb{F}_q$ such that for every word of weight 3 its nonzero positions form a block of $\mathcal{S}$ and all blocks of $\mathcal{S}$ are of this form? We refer to Assmus and Mattson (1971).

One easily sees that the binary Hamming codes as defined in Chapter 9 have the property that the words of weight 3 represent a Steiner triple system on $2^m - 1$ points

(Chapter 9, Exercise 8). In fact if a Steiner triple system on n points corresponds to the set of words of weight 3 of a binary linear code C, then $n = 2^m - 1$ and C is the Hamming code and the Steiner system is $\mathrm{PG}(m-1, 2)$.

For the Steiner triple systems on nine, respectively thirteen points, the problem has been solved (cf. Driessen *et al.* (1976)). Let $q \equiv 1 \pmod 3$ and let α be a primitive cube root of unity in $\mathbb{F}_q$. Define

$$H := \begin{pmatrix} 0 & 0 & 0 & -1 & -\alpha & -\alpha^2 & 1 & 1 & 1 \\ 1 & 1 & 1 & 0 & 0 & 0 & -1 & -\alpha & -\alpha^2 \\ -1 & -\alpha & -\alpha^2 & 1 & 1 & 1 & 0 & 0 & 0 \end{pmatrix}.$$

Then H is the parity check matrix of a [9,6] linear code over $\mathbb{F}_q$ such that the nonzero positions of codewords of weight 3 represent $\mathrm{AG}(2,3)$.

There is a geometric way of viewing this construction. The nine columns of the matrix H span nine points in $\mathrm{PG}(2,q)$ which form the inflection points of a non-singular cubic curve (whose equation is $x_1^3 + x_2^3 + x_3^3 - 3cx_1x_2x_3 = 0$, for any c with $c^3 \neq 1$). Any line containing two of these inflection points contains the third also. Classically, these nine points are known as the *Hessian configuration*. In the special case $q = 4$, when $x^2 = \overline{x}$, we can choose $c = 0$, and the equation becomes $x_1\overline{x_1} + x_2\overline{x_2} + x_3\overline{x_3} = 0$; the nine points form a *unital* in $\mathrm{PG}(2,4)$ (see Chapter 1). The code is associated with the Hessian configuration by the correspondence principle of Chapter 9, Exercise 12.

We refer to Driessen *et al.* (1976) for a proof that the two nonisomorphic Steiner triple systems on thirteen points do not arise from linear codes in this way. See also Doyen *et al.* (1978).

In our derivation of Theorem (10.4) we have demonstrated that the requirement that all the codewords have α as a zero corresponds to prescribing the parity check matrix $H := (1 \quad \alpha \quad \alpha^2 \quad \dots \quad \alpha^{n-1})$, where we have to interpret α^j as $(\epsilon_0, \epsilon_1, \dots, \epsilon_{m-1})^{\top}$ if $\alpha^j = \sum_{i=0}^{n-1} \epsilon_i \alpha^i$.

A second useful example is obtained by taking $g(x) := (x-1)m_1(x)$. Clearly the new code is a subcode of the Hamming code. The requirement that 1 is a zero of all the codewords means that they all have even weight. So we now have an $[n, n-m-1]$ code consisting of all the words of even weight in the $[n, n-m]$ Hamming code. In terms of parity check matrices we can say that $\mathbf{1}$ has been added as a parity check.

We now introduce a very important class of codes known as *BCH codes.* They were discovered by R. C. Bose, D. K. Ray-Chaudhuri, and A. Hocquenghem (and therefore should actually be called BR-CH codes).

(10.5) DEFINITION. A cyclic code of length n over $\mathbb{F}_q$ is called a *BCH code of designed distance* δ if its generator $g(x)$ is the least common multiple of the minimal polynomials of $\beta^l, \beta^{l+1}, \dots, \beta^{l+\delta-2}$ for some l, where β is a primitive n-th root of unity in some extension field of $\mathbb{F}_q$. Usually we take $l = 1$ (so-called *narrow sense* BCH codes). If $n = q^m - 1$, that is, β is a primitive element of $\mathbb{F}_{q^m}$, then the BCH code is called *primitive.*

Note that Hamming codes are primitive BCH codes with designed distance 3.

The terminology "designed distance" is explained by the following theorem. (Without loss of generality we give the proof for narrow sense BCH codes.)

(10.6) Theorem. *The minimum distance of a BCH code with designed distance δ is at least δ.*

PROOF. We use the abbreviated notation introduced above and define the $m(\delta - 1)$ by n matrix H over $\mathbb{F}_q$ by

$$H := \begin{pmatrix} 1 & \beta & \beta^2 & \dots & \beta^{n-1} \\ 1 & \beta^2 & \beta^4 & \dots & \beta^{2n-2} \\ \vdots & & & & \vdots \\ 1 & \beta^{\delta-1} & \beta^{2(\delta-1)} & \dots & \beta^{(n-1)(\delta-1)} \end{pmatrix}.$$

A word $\mathbf{c}$ is in the BCH code iff $\mathbf{c}H^\top = \mathbf{0}$. If $\mathbf{c}$ had weight $< \delta$, then there would be $\delta - 1$ columns of H that are linearly dependent over $\mathbb{F}_q$. The determinant of the submatrix of H obtained by taking these columns would have the form

$$\begin{vmatrix} \xi_1 & \xi_2 & \dots & \xi_{\delta-1} \\ \xi_1^2 & \xi_2^2 & \dots & \xi_{\delta-1}^2 \\ \vdots & & & \vdots \\ \xi_1^{\delta-1} & \xi_2^{\delta-1} & \dots & \xi_{\delta-1}^{\delta-1} \end{vmatrix} = \xi_1\xi_2\dots\xi_{\delta-1}\prod_{i>j}(\xi_i - \xi_j) \neq 0.$$

This contradiction shows that all codewords $\mathbf{c} \neq \mathbf{0}$ indeed have weight at least δ. □

Let us now take a look at BCH codes from the point of view of the group theorist. We consider a primitive BCH code of length $n = q^m - 1$ over $\mathbb{F}_q$ with prescribed zeros $\alpha, \alpha^2, \dots, \alpha^{d-1}$, where α is a primitive element in $\mathbb{F}_{q^m}$. We now denote the *positions* in codewords by the nonzero field elements $X_i := \alpha^i$ $(i = 0, 1, \dots, n-1)$. We extend the code by adjoining an overall parity check. The additional position is denoted by the zero of the field (also as X_∞ by some authors). We shall show that the extended BCH code is invariant under the permutations of the *affine permutation group* on $\mathbb{F}_{q^m}$ acting on the positions of codewords. This group, which we denote by $\mathrm{AGL}(1, q^m)$, consists of the permutations

$$P_{u,v}(X) := uX + v \qquad (u, v \in \mathbb{F}_{q^m}, \quad u \neq 0).$$

This is a doubly transitive group.

We first observe that $P_{\alpha,0}$ is the cyclic shift acting on the positions of the BCH code and that it leaves the parity check fixed. Therefore $P_{\alpha,0}$ transforms the extended

BCH code into itself. Let $(c_0, c_1, \ldots, c_{n-1}, c_\infty)$ be any codeword in the extended BCH code and apply $P_{u,v}$ to the positions to obtain the word $(c'_0, c'_1, \ldots, c'_\infty)$. To show that we again have a codeword in the extended code we must prove that for $1 \le k \le d-1$ we have $\sum_{i=0}^{n-1} c'_i \alpha^{ik} = 0$. This sum is

$$\begin{aligned}\sum_{i=0}^{n-1} c_i(u\alpha^i + v)^k + c_\infty v^k &= \sum_{i=0}^{n-1} c_i \sum_{l=0}^{k} \binom{k}{l} u^l \alpha^{il} v^{k-l} + c_\infty v^k \\ &= \sum_{i=0}^{n-1} c_i \sum_{l=1}^{k} \binom{k}{l} u^l \alpha^{il} v^{k-l} \\ &= \sum_{l=1}^{k} \binom{k}{l} u^l v^{k-l} \sum_{i=0}^{n-1} c_i (\alpha^l)^i = 0\end{aligned}$$

because the inner sum is 0 for $l = 1, 2, \ldots, d-1$.

We have proved the following theorem.

(10.7) Theorem. *Every extended primitive BCH code of length $n+1 = q^m$ over $\mathbb{F}_q$ is invariant under the affine permutation group acting on $\mathbb{F}_{q^m}$.* □

(10.8) Corollary. *The minimum weight of a primitive binary BCH code is odd.*

PROOF. Let C be such a code. We have shown that the automorphism group of the extended code $\overline{C}$ is transitive on the positions. The same therefore holds for the subset of words of minimum weight. Hence $\overline{C}$ has a word $\mathbf{c}$ of minimum weight with $c_\infty = 1$. Since all weights in $\overline{C}$ are even, this proves the assertion. □

The argument shows that any binary code, for which the extended code is invariant under a transitive permutation group, has odd minimum weight.

A special example of BCH codes are the BCH codes with $n = q-1$. These codes are called *Reed Solomon codes* (RS codes). In the narrow sense case we then have

$$g(x) = \prod_{i=1}^{n-k} (x - \alpha^i),$$

where α is a primitive element of the alphabet $\mathbb{F}_q$. By Theorem (10.6), the code has distance $d \ge n-k+1$, and k is the dimension. By the Singleton bound (Chapter 9, Exercise 3), d cannot be larger. So these codes are MDS codes!

Exercises

1. Let $n := (q^m - 1)/(q-1)$. Suppose that $(m, q-1) = 1$. Show that the $[n, n-m]$ Hamming code over $\mathbb{F}_q$ is cyclic.

2. Let C be a ternary [9,3] code with the property

$$\forall_{(c_0, c_1, \ldots, c_8) \in C} [(-c_8, c_0, \ldots, c_7) \in C].$$

Determine a generator matrix for C.

3. Consider a binary cyclic code of length 23 with dimension k, where $1 < k < 22$. Prove that $k = 11$, or $k = 12$.

4. Let C be a binary code of length 21 with generator

$$g(x) = m_0(x)m_1(x)m_3(x)m_7(x).$$

Determine the dimension of C. Show that C is selforthogonal. Show that C has minimum distance at least 6.

5. Let $x^{i_1} + x^{i_2} + \ldots + x^{i_k}$ be a codeword in the code of the previous exercise. Then k is even. Show that if $d(x) = x^{21-i_1} + \ldots + x^{21-i_k}$, then $c(x)d(x) = 0$ in $\mathbb{F}_2[x]/(x^{21} - 1)$. Now prove that C has minimum distance 8. (Hint: prove that $k(k-1) \equiv 0 \pmod 4$.)

6. Let C be an RS code over $\mathbb{F}_q$. Prove that the extended code is also an MDS code.

7. Let C be an $[n, k, d]$ RS code ($n = q - 1$, $d = n - k + 1$). Define the code C' of length q by adjoining $-c(\alpha^d)$ as an extra symbol. Show that C' is an MDS code. Now show that from the RS code C we can construct an $[n + 2, k, d + 2]$ MDS code.

8. Let $\overline{C}$ be the [8,4,5] extended RS code over $\mathbb{F}_8$. Show that $\overline{C}$ is selfdual. Let α be a primitive element of $\mathbb{F}_8$ with $\alpha^3 + \alpha + 1 = 0$. The normal basis $\{\alpha, \alpha^2, \alpha^4\}$ has the property that $\mathrm{Tr}(\alpha_i\alpha_j) = \delta_{ij}$ for any two basis elements α_i and α_j. Map $\mathbb{F}_8$ to $\mathbb{F}_2^3$ using the representation over this basis. Show that the resulting code $\mathcal{G}$ is a [24,12] selfdual code. By inspection one can check that the basis vectors of $\mathcal{G}$ have weight 8. Now show that $\mathcal{G}$ has minimum distance 8. (This construction of the extended binary Golay code, the topic of the next chapter, is due to Pasquier (1980).)

9. Consider an $[n, 3]$ Reed Solomon code over $\mathbb{F}_q$ ($n = q - 1$). Show that the set of points of $\mathrm{PG}(2, q)$ associated with it as in Chapter 9, Exercise 12 (spanned by the columns of the generator matrix) is an n-arc (that is, no three collinear).

11. The Golay codes

In this chapter we treat two remarkable codes that were discovered by M. J. E. Golay (1949). Both codes are perfect, both have many relations to design theory, and also both have a remarkable automorphism group. We start with the *binary Golay code* $\mathcal{G}_{23}$. We shall mention several constructions for this code. The fact that the code is unique will follow from the uniqueness proof for the extended code $\mathcal{G}_{24}$ and the transitivity of the automorphism group of $\mathcal{G}_{24}$.

The binary Golay code is a [23,12,7] code. Since these parameters imply equality in the sphere-packing bound (Chapter 9, Exercise 1), this code is perfect. (It is in fact the only nontrivial perfect e-error-correcting code with $e > 2$.) In the extended code all weights are divisible by 4; we call such a code *doubly even*. (Note that if the basis vectors of a selfdual code all have weight divisible by 4, then the code is doubly even: this follows, by an easy induction, from the fact that the sum of two orthogonal vectors each with weight divisible by 4 has weight divisible by 4.) We now establish the uniqueness of this extended code and in the proof also provide a construction.

(11.1) Theorem. *Let C be a binary code of length 24 with minimum distance 8, and suppose that $\mathbf{0} \in C$ and $|C| = 2^{12}$. Then C is determined up to equivalence. The code is called $\mathcal{G}_{24}$. (So: the extended binary Golay code is unique).*

PROOF. (a) Puncturing on any position leads to a $(23, 2^{12}, 7)$ code. Since such a code is perfect, its weight enumerator is determined (cf. Chapter 9, Exercise 9). In fact $A_0 = A_{23} = 1$, $A_7 = A_{16} = 253$, $A_8 = A_{15} = 506$, $A_{11} = A_{12} = 1288$. This immediately implies that the code C only has words of weight 0,8,12,16, and 24. However, the same is true for the code $C + \mathbf{c}$ for any $\mathbf{c} \in C$. Therefore not only all the weights in C are divisible by 4 but also all the distances between codewords are divisible by 4. This implies that $\langle \mathbf{c}, \mathbf{c}' \rangle = 0$ for any two codewords $\mathbf{c}$, $\mathbf{c}'$ in C. So, the words of C span a (doubly even) self-orthogonal code. Such a code has dimension at most 12 and thus we see that the code C was already linear!

(b) Take any codeword $\mathbf{c}$ of weight 12 as a basis vector for C. The residual code must have dimension 11 and it has only even weights. So, the residual code is the

[12,11,2] even-weight code. Therefore C has a generator matrix G of the form

$$(*) \qquad G = (I_{12} \quad P), \quad \text{where} \quad P = \begin{pmatrix} 0 & \mathbf{1} \\ \mathbf{1}^{\top} & A \end{pmatrix},$$

(so A is of size 11 by 11).

(c) Since C has minimum distance 8, every row of A has at least six 1s. From the top row of G we see that every row of A therefore must have exactly six 1's. Clearly, any two rows of A have at most three 1's in common. Using the top row of P we then see that any two rows of A have exactly three 1's in common. This forces A to be the incidence matrix of a 2-(11,6,3) design. The uniqueness of that design is well known (easily proved by hand — see Exercise 8(a) of Chapter 1). □

It is easily checked (Exercise 1) that if we indeed take A to be the incidence matrix of a 2-(11,6,3) design in (*), then G does generate a [24,12,8] code.

From the words of weight 8 of $\mathcal{G}_{24}$ one finds a 5-(24,8,1) design, that is, a Steiner system. We shall also show that only one such design exists. (This existence and uniqueness proof is the one we promised in Chapter 1, and is a good deal shorter than the proofs we outlined there!)

(11.2) Theorem. *The words of weight 8 in $\mathcal{G}_{24}$ are the characteristic functions of the blocks of a 5-$(24, 8, 1)$ design.*

PROOF. Since $\mathcal{G}_{24}$ has distance 8, two words of weight 8 cannot have five 1's in common. Hence the 759 words of weight 8 cover $759 \cdot \binom{8}{5} = \binom{24}{5}$ distinct 5-tuples from $\{1, 2, \ldots, 24\}$. □

(11.3) Theorem. *The 5-$(24, 8, 1)$ Steiner system is unique.*

PROOF. (a) Let S be such a system. We saw in Chapter 1 (using intersection triangles) that two blocks of S meet in 0, 2 or 4 points. Thus, the code C spanned by the blocks of S is self-orthogonal and doubly even. To see that C has minimum distance 8, observe that $C^{\perp}$ must have minimum distance at least 6. This follows from the fact that the blocks of S assume all possible 0-1 configurations on a given 5-set of points (cf. Chapter 1).

(b) Fix three points. The derived design with respect to these points is a 2-(21,5,1) design, that is, the plane $PG(2, 4)$. In Chapter 13 we shall see that the rows of the incidence matrix of $PG(2, 4)$ span a code of dimension 10. This implies that C has dimension 12 and by Theorem (11.1) we are done. □

As observed above, Theorem (11.1) provides one construction of $\mathcal{G}_{24}$. Note that from the construction of Paley matrices in Chapter 1 we see that the 2-(11,6,3) design (and hence also $\mathcal{G}_{24}$) has an automorphism of order 11.

(11.4) Proposition. *Let N be the 12 by 12 adjacency matrix of the graph formed by the vertices and edges of the icosahedron. Then*

$$G := (I_{12} \quad J - N)$$

is the generator matrix of $\mathcal{G}_{24}$.

PROOF. Two antipodal vertices of the icosahedron have no common neighbour. Any two other vertices have two common neighbours. This already implies that G generates a [24,12] selfdual doubly even code C. Since $J-N$ is symmetric, the matrix $(J - N \; I_{12})$ also generates C (because $C = C^{\perp}$). So, if there were a word of weight 4, then there would have to be two rows of G whose sum has weight 4, which is clearly false. By Theorem (11.1) we are done. □

The argument of the proof shows that $\mathcal{G}_{24}$ has an automorphism which interchanges the first twelve coordinates with the second twelve. This, and the fact that the automorphism group of the icosahedron is transitive on the twelve vertices, shows that the automorphism group of $\mathcal{G}_{24}$ is transitive. So if we puncture on any position, we always find the same code (up to equivalence). This shows that $\mathcal{G}_{23}$ is unique.

We present a construction of the Golay code due to R. J. Turyn. We consider the [7,4] Hamming code in the representation of the example following (10.4). Call this code $\mathcal{H}$. Let $\mathcal{H}^*$ be the code obtained by reversing the order of the symbols in $\mathcal{H}$. By inspection we see that the codes $\overline{\mathcal{H}}$ and $\overline{\mathcal{H}^*}$ are [8,4] selfdual codes with the property $\overline{\mathcal{H}} \cap \overline{\mathcal{H}^*} = \{\mathbf{0}, \mathbf{1}\}$.

(11.5) Proposition. *Let C be the $[24, 12]$ binary code defined by*

$$C := \{(\mathbf{a}+\mathbf{x}; \mathbf{b}+\mathbf{x}; \mathbf{a}+\mathbf{b}+\mathbf{x}) : \mathbf{a} \in \overline{\mathcal{H}}, \mathbf{b} \in \overline{\mathcal{H}}, \mathbf{x} \in \overline{\mathcal{H}^*}\}.$$

Then C is $\mathcal{G}_{24}$.

PROOF. Clearly the words $(\mathbf{a}; \mathbf{0}; \mathbf{a})$, $(\mathbf{0}; \mathbf{b}; \mathbf{b})$, and $(\mathbf{x}; \mathbf{x}; \mathbf{x})$ where $\mathbf{a}$ and $\mathbf{b}$ run through a basis of $\overline{\mathcal{H}}$ and $\mathbf{x}$ runs through a basis of $\overline{\mathcal{H}^*}$, form a basis of C. So C is indeed a [24,12] code. Since $\overline{\mathcal{H}}$ and $\overline{\mathcal{H}^*}$ are selfdual, any two basis vectors have inner product 0. So C is selfdual. Since all basis vectors have weight $\equiv 0 \pmod 4$, we conclude that C is doubly even. Clearly the three subwords of length 8 in a codeword have even weight. Suppose a codeword $\mathbf{c}$ has weight ≤ 4. Then one of the subwords is $\mathbf{0}$ and this implies that $\mathbf{x}$ is $\mathbf{0}$ or $\mathbf{1}$. Since $\mathbf{a}$, $\mathbf{b}$, and $\mathbf{a}+\mathbf{b}$ all have weight 0, 4, or 8, we conclude that $\mathbf{c} = \mathbf{0}$. So C is a [24,12,8] code, that is, $\mathcal{G}_{24}$. □

In Chapter 14 we shall see that the (essentially unique) [23,12] binary cyclic code (cf. Chapter 10, Exercise 3) is $\mathcal{G}_{23}$.

A remarkably easy construction (that has a rather complicated proof that it works) is obtained as follows. Start with $\mathbf{0} \in \mathbb{F}_2^{24}$ and make a code of length 24 by successively taking the lexicographically least word that has not been used and that has distance ≥ 8 to the already chosen set; (so $(0,0,\ldots,0,1,1,\ldots,1)$ of weight 8 is the second word). This yields $\mathcal{G}_{24}$ after 2^{12} choices.

In Chapter 10, Exercise 8 we saw a construction of $\mathcal{G}_{24}$ from a RS code over $\mathbb{F}_8$. The next construction is probably the nicest. It is due to J. H. Conway. We first define a code $\mathcal{H}_6$ over the field $\mathbb{F}_4$, known as the *hexacode*. The alphabet is represented as $\{0,1,\omega,\overline{\omega}\}$, where $\overline{\omega}=\omega+1=\omega^2$.

(11.6) DEFINITION. The code $\mathcal{H}_6$ is the [6, 3] code over $\mathbb{F}_4$ generated by the matrix

$$G:=\begin{pmatrix}1&0&0&1&\overline{\omega}&\omega\\0&1&0&1&\omega&\overline{\omega}\\0&0&1&1&1&1\end{pmatrix}.$$

We first observe that this code can also be defined by

$$\mathcal{H}:=\{(a,b,c,f(1),f(\omega),f(\overline{\omega}))|f(x):=ax^2+bx+c,(a,b,c)\in\mathbb{F}_4^3\}.$$

Note that if two of $\{a,b,c\}$ are 0, then the codeword has weight 4; if one of them is 0, then $f(x)=0$ for exactly one $x\neq 0$; and if all of them are $\neq 0$, then f has two such zeros or none. So a code word in $\mathcal{H}_6$ has weight 0, 4, or 6.

REMARK. This can also be seen as follows. The 1-dimensional subspaces of $\mathbb{F}_4^3$ spanned by the columns of G are the points of a conic in PG(2, 4) together with its nucleus (see Chapter 1 — the conic has equation $x_1x_3=x_2^2$, and its nucleus is [0, 1, 0]). These points form a (Type II) oval in PG(2, 4), meeting every line in 0 or 2 points. Thus the assertion follows from Chapter 9, Exercise 12.

(11.7) DEFINITION. We define a binary code C as follows. The words are represented by 4 by 6 binary matrices M (with rows $\mathbf{m}_0,\mathbf{m}_1,\mathbf{m}_2,\mathbf{m}_3$). A matrix M represents a codeword if and only if
(1) every column of M has the same parity as $\mathbf{m}_0$;
(2) $\mathbf{m}_1+\omega\mathbf{m}_2+\overline{\omega}\mathbf{m}_3\in\mathcal{H}_6$.

(11.8) Theorem. *The code C defined in (11.7) is $\mathcal{G}_{24}$.*

PROOF. (a) Conditions (1) and (2) are clearly linear. We can choose the parity in two ways and a codeword $\mathbf{c}\in\mathcal{H}_6$ in 4^3 ways. Every coordinate c_i of $\mathbf{c}$ corresponds to two (complementary) possible columns of M. So, for the first five columns we have a total of 2^5 choices and then the condition on the parity of the first row uniquely determines the sixth column. Therefore $|C|=2\cdot 4^3\cdot 2^5=2^{12}$, that is, C is a [24,12] code.

(b) If the chosen parity is *even* and $\mathbf{c}\neq\mathbf{0}$, then there are at least four columns in M with weight ≥ 2. So the matrix has weight at least 8. If the parity is even and

$\mathbf{c} = \mathbf{0}$ but the matrix is not all-zero, then at least two columns of M have four 1's. Again the weight is at least 8.

(c) If the chosen parity is *odd*, then to show that the weight is again at least 8, we only have to show that it is impossible that every column of M has exactly one 1. If this were so, then the fact that $\mathbf{c}$ has even weight would force $\mathbf{m}_0$ to have even weight, a contradiction.

We have shown that C is a [24,12,8] code and hence we are done. □

We now come to the *ternary Golay code* $\mathcal{G}_{11}$ and its extension $\mathcal{G}_{12}$. The code $\mathcal{G}_{11}$ is a ternary [11,6,5] code. Again, these parameters imply that the code is perfect. It has been shown that if the alphabet is a field, then there is no other nontrivial perfect 2-error-correcting code. It has also been shown that any $(11, 3^6, 5)$ code is equivalent to $\mathcal{G}_{11}$ (but this is much more difficult than in the case of $\mathcal{G}_{24}$).

In Chapter 14 we shall see a construction of the ternary Golay code as a quadratic residue code. Here we give a few combinatorial constructions. Let S_5 be the Paley matrix of size 5 as defined in Chapter 1, that is,

$$S_5 := \begin{pmatrix} 0 & + & - & - & + \\ + & 0 & + & - & - \\ - & + & 0 & + & - \\ - & - & + & 0 & + \\ + & - & - & + & 0 \end{pmatrix},$$

where + indicates 1 and − indicates −1. Consider the [11,6] ternary code C defined by the generator matrix $G := (I_6 \quad P)$, where P is the matrix S_5 bordered on top by $\mathbf{1}$ of length 5. The fact that S_5 satisfies $S_5 S_5^\top = 5I - J$ immediately shows that $\overline{C}$ is a [12,6] selfdual code. Therefore all the words of $\overline{C}$ have weight divisible by 3. The generator $\overline{G}$ of the extended code is obtained by adjoining the column $(0, -1, -1, -1, -1, -1)^\top$ to G. Every row of $\overline{G}$ has weight 6 and since a linear combination of two rows has weight at least $2 + 2$, it has weight 6. Therefore such a combination has exactly two zeros among the last six coordinates and this forces a linear combination of three rows to have weight at least $3 + 1$, and hence again at least 6. This shows that $\overline{C}$ has minimum distance 6. We take this as definition of $\mathcal{G}_{12}$. If we use the uniqueness result mentioned above we have the following proposition.

(11.9) Proposition. *The code C defined above is $\mathcal{G}_{11}$.* □

In the next construction we use the [4,2] ternary Hamming code. Let H be a 2 by 4 parity check matrix of this code. The matrices I and J used below, have size 4 by 4.

(11.10) Proposition. *The matrix G given by*

$$G := \begin{pmatrix} J+I & I & I \\ O & H & -H \end{pmatrix}$$

generates a $[12,6,6]$ *selfdual code (and by uniqueness this is* $\mathcal{G}_{12}$*).*

PROOF. Let C be the code generated by G. Since $J+I$ has rank 4, C has dimension 6. Since the ternary Hamming code is selfdual, C is also selfdual. It suffices to show that weight 3 cannot occur. A linear combination of the first four rows obviously has weight > 4 and a linear combination of the last two has weight 6 by definition. The reader should now convince himself that a linear combination involving both kinds of rows cannot have weight 3 (again because $J+I$ has full rank). □

In the following construction we start with a code C over $\mathbb{F}_9$. Let α be a primitive element of $\mathbb{F}_9$, $i := \alpha^2$. Then $i^2 = \alpha^4 = -1$. We define $\overline{\xi} := \xi^3$ for $\xi \in \mathbb{F}_9$ and write elements of this field as $a+bi$ with a and b in $\mathbb{F}_3$. Then $\overline{a+bi} = a-bi$. We extend the notation to codewords $\mathbf{c}$ by defining $\overline{\mathbf{c}} = (\overline{c_1}, \dots, \overline{c_n})$ if $\mathbf{c} = (c_1, \dots, c_n)$. We call the code C *conjugate dual* if $\mathbf{c} \to \overline{\mathbf{c}}$ maps C to $C^\perp$. We "project" C to a code of length $2n$ by $a+bi \to (a,b)$. Since $(a+bi)(c-di) = (ac+bd)+(bc-ad)i$, the projection of a conjugate dual code of length n over $\mathbb{F}_9$ is a selfdual code of length $2n$ over $\mathbb{F}_3$.

(11.11) Proposition. *Let* C *be the* $[6,3]$ *code over* $\mathbb{F}_9$ *defined by the generator matrix* $G := (I \quad \alpha A)$*, where* $A := I + iJ$*. Then the projection of* C *is a* $[12,6,6]$ *ternary code (and therefore* $\mathcal{G}_{12}$*).*

PROOF. (a) From $A\overline{A}^{\mathsf{T}} = I$ we find $G\overline{G}^{\mathsf{T}} = O$, that is, C is conjugate dual.

(b) This implies that both G and $(-\alpha\overline{A} \quad I)$ generate C. This shows that C has minimum distance 4.

(c) Since the projection of C is selfdual, it must have minimum distance at least 6 and we are done. □

(This proof is due to J. I. Hall; a similar construction was given by D. Y. Goldberg (1986).)

To show a connection with a previous topic, we now describe a construction of a strongly regular graph using $\mathcal{G}_{11}$. We partition the space $\mathbb{F}_3^{11}$ into 3^5 cosets of the ternary Golay code. A coset is represented by its word of minimum weight, so by $\mathbf{0}$, one of the 22 words of weight 1, or one of the 220 words of weight 2. The weight of this so-called coset leader is called the weight of the coset.

(11.12) Proposition. *There exists a strongly regular graph with parameters* $(243, 22, 1, 2)$.

PROOF. Take the cosets of $\mathcal{G}_{11}$ as vertices and join two vertices by an edge iff the difference of the corresponding cosets is a coset of weight 1. The valency is clearly 22. We claim the every edge is in a unique triangle and every nonedge is in a unique quadrangle. To prove this we may assume (by linearity) that one of the vertices has weight 0. An adjacent vertex corresponds to a coset with a leader **a** of weight 1 and the only vertex joined to both of them has 2**a** as leader because $\mathcal{G}_{11}$ has minimum distance 5. For the same reason a coset adjacent to **0** and the vertex represented by $(a, b, 0, \dots, 0)$ must be represented by $(a, 0, \dots, 0)$ or by $(0, b, 0, \dots, 0)$. □

Exercises

1. Show that if in (*) in the proof of Theorem (11.1) we take A to be the incidence matrix of a 2-(11,6,3) design, then G indeed generates a [24,12,8] code.

2. From (11.5) we see that $\mathcal{G}_{24}$ has a subcode with 32 words that have a 0 in the first eight positions. Show that the same is true if we require that $c_8 = 1$ and exactly one of $c_1, \dots, c_7$ is 1. Prove that if we take the union of these eight subcodes and then delete the first eight symbols of the words, we find a $(16, 256, 6)$ code. This code is known as the *Nordstrom–Robinson* code. It is known that a $(16, M, 6)$ code must have $M \leq 256$.

3. Show that a $(12, 32, 5)$ code exists (see Chapter 9, Exercise 4).

4. In (11.7), consider all codewords of the form $(A\ A\ A)$, where A is a 4 by 2 matrix. Show that the corresponding matrices A form the [8,4] extended Hamming code.

5. Consider the set P of all words in $\mathcal{G}_{24}$ for which the representation of (11.7) has the form $M = (B \quad A)$, where B is 4 by 2 with first column $(*111)^{\top}$, the $*$ and the second column of B arbitrary. We index the rows and columns of A with $\mathbb{F}_4$ as follows: row number 0, 1, 2, 3 corresponds to $x = 0, 1, \omega, \overline{\omega}$ respectively; and similarly for the columns and the coordinate y. We consider A as $AG(2,4)$. Show that a word of weight 8 in P has five 1's or one 1 in the remaining five positions of B. Show that a word in P with four 1's in the first column of B is described by $y =$ constant. Show that a word in P that has one 1 in the second column of B, corresponding to $x = \xi$, is described by $x = \xi y + \eta$. This shows that these coordinates describe these words as the lines of $PG(2,4)$, the word with eight 1's in B being the line at infinity. This should not be surprising since (by (11.2)) we are looking at a derived design of $S(5,8,24)$ which is a 2-(21,5,1) design; this design is unique, namely $PG(2,4)$.

6. (a) Show that the cyclic permutation $(\mathbf{m}_1, \mathbf{m}_2, \mathbf{m}_3)$ of M in (11.7) leaves $\mathcal{G}_{24}$ invariant.

(b) Show that interchanging the last two rows and the last two columns of M leaves $\mathcal{G}_{24}$ invariant (consider $\omega \leftrightarrow \overline{\omega}$).

(c) The permutation (12)(34) leaves $\mathcal{H}_6$ invariant. Prove this, and find the corresponding permutation of $\mathcal{G}_{24}$.

7. Find all binary perfect 3-error-correcting codes.

8. Let C be a ternary selfdual [12,6,6] code. Using only this assumption, prove that up to a monomial transformation $C = \mathcal{G}_{12}$.

9. Let $C = \mathcal{G}_{11}^{\perp}$, and define a graph with vertex set C by declaring two codewords to be adjacent if their distance is 6. Prove that the graph is strongly regular, with parameters (243, 132, 81, 60).

10.(a) Show that, if two blocks of the 5-(24, 8, 1) design meet in four points, then their symmetric difference is a block.

(b) Now let $\mathcal{D} = (X, \mathcal{B})$ be a $(t+1)$-$(v, 2t, 1)$ design, for some t, with $v > 2t$, which has the property

(*) *if two blocks of $\mathcal{D}$ meet in t points, then their symmetric difference is a block.*

Choose a point $\infty \in X$, and let $Y = X \setminus \{\infty\}$. Let Ω be the set of all subsets of Y of cardinality at most $t-1$, and define a graph on Ω, whose edges are labelled with elements of Y, as follows:

(a) for each set $A \subset Y$ with $|A| \leq t-1$ and $y \in A$, join A to $A \setminus \{y\}$ by an edge labelled y;

(b) for each pair A_1, A_2 of $(t-1)$-subsets of Y such that $A_1 \cup A_2 \cup \{\infty, y\} \in \mathcal{B}$, join A_1 to A_2 by an edge labelled y.

Prove that each vertex lies on a unique edge with any given label.

Now take $\mathsf{F} = \mathsf{F}_2$, and let $V = \mathsf{F}^{v-1}$, with its basis labelled by elements of Y. Define a map $\phi : V \to \Omega$ as follows: given $\mathbf{v} \in V$, take the path starting at $\emptyset \in \Omega$ and following those edges whose labels index non-zero coordinates of $\mathbf{v}$; then $\phi(\mathbf{v})$ is the other end of this path. Prove that ϕ is well-defined (that is, the endpoint does not depend on the order in which the labels are used). Show further that

$$\{\mathbf{v} \in V : \phi(\mathbf{v}) = \emptyset\}$$

is a linear perfect $(t-1)$-error-correcting code.

Deduce that $\mathcal{D}$ is either a 3-$(2^n, 4, 1)$ design whose blocks are the planes in AG$(n, 2)$, or the unique 5-(24, 8, 1) design.

11. Let Γ be a graph in which every edge lies in a unique triangle, and every non-edge in a unique quadrangle.

(a) Prove that Γ is regular, with valency k (say) — thus Γ is strongly regular (or complete).

(b) Show that $k = 2$, 4, 14, 22, 112 or 994.

(For $k = 2, 4$ the graph is a triangle or $L_2(3)$ respectively; we saw an example with $k = 22$ in (11.12). The other cases are undecided.)

(c) Now let C be any ternary linear perfect 2-error-correcting code. Show that the construction of (11.12), applied to C, gives a strongly regular graph with $\lambda = 1$ and $\mu = 2$. Deduce that any such code must have length 11.

(d) Can you extend this argument to other finite fields?

12. Reed–Muller and Kerdock codes

In this chapter, we discuss a class of binary codes with close connections to finite geometries. These codes were first treated by D. E. Muller (1954) and I. S. Reed (1954). The codes are called *Reed–Muller codes*. There are several ways to describe the codes. In one of them, the codewords are linear combinations of characteristic functions of certain affine subspaces in $\mathrm{AG}(m,2)$. We introduce the codes using another description.

We consider Boolean functions in m variables $x_1, \ldots, x_m$. Such a function is a polynomial in these variables and each term has the form $x_{i_1}x_{i_2}\cdots x_{i_k}$ with $k \leq m$. We let the variables correspond to the coordinates of points in $\mathbb{F}_2^m$ (a model of $\mathrm{AG}(m,2)$). Since we need the notation later, we number the points in this space as P_0 to P_{n-1}, where $n = 2^m$ and P_i has coordinates $\xi_{i1}, \ldots, \xi_{im}$ if $i = \sum_{j=1}^{m} \xi_{ij} 2^{j-1}$. Using this numbering of the points in $\mathbb{F}_{2^m}$, we identify a Boolean function with its list of values, and interpret this list of values as a word in the space $\mathbb{F}_2^n$. For example, the Boolean function x_1 of degree 1 takes the value 1 in all points P_i of $\mathbb{F}_{2^m}$ for which $\xi_{i1} = 1$, that is, the points with an odd index. So this function corresponds to the word $(0,1,0,1,\ldots,0,1)$ of length 2^m.

(12.1) DEFINITION. The r-th order *Reed–Muller code* of length $n = 2^m$, denoted by $\mathcal{R}(r,m)$, has as codewords (the lists of values of) all the Boolean functions of degree at most r.

From now on we call these codes RM codes.

(12.2) Theorem. *$\mathcal{R}(r,m)$ has minimum distance 2^{m-r}.*

PROOF. The proof is by induction on m. For $m = r$, the code is the whole space $\mathbb{F}_2^m$, and obviously has minimum distance 1. Now consider the step from m to $m+1$. Take an arbitrary function $\mathbf{w} \neq \mathbf{0}$ and write it as $\mathbf{w} = \mathbf{u}' + \mathbf{v}'$, where $\mathbf{u}'$ and $\mathbf{v}'$ are sums of monomials not involving, resp. involving, the variable x_{m+1}. Then we have $\mathbf{u}' = (\mathbf{u},\mathbf{u})$ and $\mathbf{v}' = (\mathbf{0},\mathbf{v})$, where $\mathbf{u} \in \mathcal{R}(r,m)$ and $\mathbf{v} \in \mathcal{R}(r-1,m)$. If $\mathbf{v} = \mathbf{0}$ then $\mathbf{w} = (\mathbf{u},\mathbf{u})$ has weight at least $2 \cdot 2^{m-r}$, by the induction hypothesis. If $\mathbf{u} = \mathbf{v}$, then $\mathbf{u} \in \mathcal{R}(r-1,m)$, and so $\mathbf{w} = (\mathbf{u},\mathbf{0})$ has weight at least $2^{m-(r-1)}$. Otherwise, $\mathbf{u}$ and $\mathbf{u}+\mathbf{v}$ each have weight at least 2^{m-r}. □

We observe that the polynomial

$$\prod_{j=1}^{m}(1+\xi_{ij}+x_j)$$

is 0 everywhere except in the point P_i. This shows that the n monomials in x_1 to x_m form a basis for the space of Boolean functions on $\mathbb{F}_{2^m}$. The following theorem is now also easy.

(12.3) Theorem. *The dual of $\mathcal{R}(r,m)$ is $\mathcal{R}(m-r-1,m)$.*

PROOF. From the previous observation we see that the two codes have dimension $1+\binom{m}{1}+\cdots+\binom{m}{r}$, respectively $1+\binom{m}{1}+\cdots+\binom{m}{m-r-1}$, and the sum of these dimensions equals the word length. Each of the codes has monomials as basis vectors and the product of two such monomials has degree at most $m-1$, that is, it does not involve all the variables. This implies that the product of the two monomials is 1 in an even number of points, in other words the two basis vectors have inner product 0. $\square$

(12.4) Corollary. *The $(m-2)$-nd order RM code of length $n=2^m$ is the $[n,n-m-1]$ extended Hamming code.*

We reformulate the definition in terms of the geometry $\mathrm{AG}(m,2)$. Note that the list of values of the function x_i is the characteristic function of the $(m-1)$-dimensional affine subspace (an $(m-1)$-flat)

$$A_i := \{\mathbf{x}\in\mathbb{F}_{2^m} : x_i = 1\}.$$

The vector $\mathbf{1}$ is the characteristic function of the whole space and hence the function $1+x_i$ represents the the hyperplane parallel to A_i, that is, $\{\mathbf{x}\in\mathbb{F}_{2^m} : x_i = 0\}$. Clearly the first order RM code $\mathcal{R}(1,m)$ consists of $\mathbf{0}$, $\mathbf{1}$, and the characteristic functions of the $2^{m+1}-2$ affine subspaces of dimension $m-1$ in the space. Since the intersection of two such subspaces is empty or a flat of dimension $m-2$, the distance of any two codewords is n or $\frac{1}{2}n$. Before taking a closer look at the geometry we give an example.

Below we list the vectors corresponding to all monomials for the case $m=4$, that is, $n=16$ (including the function 1). Note that independence is obvious from the positions of the first coordinate equal to 1. Also (obvious and used in a proof above) every vector has even weight, except the last one which is the characteristic function of the set $\{P_{15}\}$.

$$
\begin{aligned}
x_0 &= (1111111111111111)\\
x_1 &= (0101010101010101)\\
x_2 &= (0011001100110011)\\
x_3 &= (0000111100001111)\\
x_4 &= (0000000011111111)\\
x_1x_2 &= (0001000100010001)\\
x_1x_3 &= (0000010100000101)\\
x_1x_4 &= (0000000001010101)\\
x_2x_3 &= (0000001100000011)\\
x_2x_4 &= (0000000000110011)\\
x_3x_4 &= (0000000000001111)\\
x_1x_2x_3 &= (0000000100000001)\\
x_1x_2x_4 &= (0000000000010001)\\
x_1x_3x_4 &= (0000000000000101)\\
x_2x_3x_4 &= (0000000000000011)\\
x_1x_2x_3x_4 &= (0000000000000001)
\end{aligned}
$$

Note (for example) that the function x_1x_3 represents the flat $A_1 \cap A_3$, consisting of the points P_5, P_7, P_{13}, and P_{15}.

(12.5) Theorem. *Let $C = \mathcal{R}(m-l, m)$ and let A be an l-flat in* AG$(m,2)$. *Then the characteristic function of A is in C.*

PROOF. Let $\mathbf{f} \in \mathbb{F}_{2^m}$ be the characteristic function of A, that is, $f_j = 1$ iff $P_j \in A$. Clearly $\mathbf{f}$ corresponds to a linear combination of the functions $x_{i_1} \cdots x_{i_k}$ (since these form a basis of all Boolean functions). We must show that in fact $\mathbf{f}$ is a linear combination of products of at most $m - l$ factors. Let α be the coefficient of the product $x_{i_1} \cdots x_{i_k}$ in the expansion of $\mathbf{f}$ and assume that $k > m - l$. Let $\mathbf{h}$ be the product of the remaining factors. Then $\alpha = \langle \mathbf{f}, \mathbf{h} \rangle$. Since $\mathbf{h}$ is the characteristic function of a flat of dimension $> m - l$, A meets this flat in an even number of points and hence $\alpha = 0$. This proves our assertion. □

This theorem and the definition show that a word is in $\mathcal{R}(r,m)$ iff it is the sum of characteristic functions of affine subspaces of AG$(m,2)$ of dimension $\geq m - r$.

We now consider the *automorphism group* of these codes, that is, the permutations of the positions that map a RM code to itself.

(12.6) Theorem. AGL$(m,2) \leq$ Aut$(\mathcal{R}(r,m))$.

PROOF. Since AGL$(m,2)$ maps k-flats to k-flats (for all k), this follows from the remark following Theorem (12.5). □

We now combine Theorem (12.3) and Theorem (12.5). Let $C = \mathcal{R}(r,m)$. We now know that every $(r+1)$-flat in $\mathrm{AG}(m,2)$ provides us with a parity check equation for C. These $(r+1)$-flats form a 3-(v,k,λ) design with

$$v = n = 2^m, \quad k = 2^{r+1}, \quad \lambda = \frac{(2^{m-2}-1)\cdots(2^{m-r}-1)}{(2^{r-2}-1)\cdots(2-1)}.$$

Consider any r-flat H in $\mathrm{AG}(m,2)$. There are $2^{m-r}-1$ distinct $(r+1)$-flats containing H. Every point outside H is in exactly one of these $(r+1)$-flats.

We describe a decoding algorithm based on these facts. Let $\mathbf{c}$ be a codeword that contains the term $x := x_{i_1}\cdots x_{i_r}$ and assume that a word $\mathbf{y}$ is received with $t < 2^{m-r-1}$ errors (see Theorem (12.2)). Now let H be the r-flat corresponding to the product of the factors x_i that do not occur in x, and let $\mathbf{h}$ be the corresponding word. Again we have $\langle \mathbf{c}, \mathbf{h}\rangle = 1$. Each of the $2^{m-r}-1$ distinct $(r+1)$-flats containing H also contains an r-flat parallel to H. If $\mathbf{h}'$ corresponds to such an r-flat, then also $\langle \mathbf{c}, \mathbf{h}'\rangle = 1$. The receiver calculates $\langle \mathbf{y}, \mathbf{h}\rangle$ and $\langle \mathbf{y}, \mathbf{h}'\rangle$ for all the parallel flats (a total of 2^{m-r} inner products). Since we have used a partitioning of the points of the space, the t errors in $\mathbf{y}$ clearly can be responsible for at most $2^{m-r}-t$ of the inner products being 0. Hence the majority of the inner products is 1 (despite the errors in $\mathbf{y}$). So, a *majority vote* decides that x was a term in $\mathbf{c}$. Once this has been done for all terms with r factors, the problem can be reduced to a RM code of order $r-1$ and we proceed in the same way. This is an example of so-called (multistep) *majority logic decoding* (easy to implement in hardware). See also Massey (1963).

Let $\mathrm{AG}(m,2)$ be considered as a representation of the additive structure of $\mathbb{F}_{2^m}$ and let α be a primitive element of this field. The mapping $x \mapsto \alpha x$ is a linear mapping of $\mathrm{AG}(m,2)$. It has order 2^m-1. By Theorem (12.5) this mapping fixes a RM code. The fact that the mapping has one orbit of length 2^m-1 shows that a RM code is (equivalent to) an *extended cyclic code*!

A cyclic definition can be given as follows, using the notation $w(j)$ for the number of 1's in the binary representation of j.

(12.7) Proposition *Let α be a primitive element in $\mathbb{F}_{2^m}$. Let $g(x) := \prod^*(x-\alpha^j)$, where the product ranges over all integers j in $(0, 2^m-2]$ with $w(j) < m-r$. If C is the cyclic code of length 2^m-1 generated by $g(x)$, then $\overline{C}$ is equivalent to $\mathcal{R}(r,m)$.*

We leave the proof as an exercise. □

We shall now introduce a class of codes that are somewhere "between" first and second order Reed–Muller codes. These codes, called *Kerdock codes* (see Kerdock (1972)), are of great combinatorial interest. We have seen that $\mathcal{R}(2,m)$ (of dimension

$1+m+\binom{m}{2})$ corresponds to the Boolean functions of degree ≤ 2 on $\mathbb{F}_{2^m}$. So $\mathcal{R}(2,m)$ is a union of cosets of $\mathcal{R}(1,m)$, each coset corresponding to some quadratic form

$$Q(\mathbf{x}) := \sum_{1\leq i<j\leq m} q_{ij}x_ix_j.$$

Corresponding to Q we have the alternating bilinear form

$$B(\mathbf{x},\mathbf{y}) := Q(\mathbf{x}+\mathbf{y}) - Q(\mathbf{x}) - Q(\mathbf{y}) = \mathbf{x}B\mathbf{y}^\top,$$

where B is a symplectic matrix (zero diagonal and $B = -B^\top$). Note that Q can be described by the part of B above the diagonal.

By Dickson's theorem (cf. Dickson (1958)) Q can be transformed by an affine transformation into the form

$$\sum_{i=1}^{h} x_{2i-1}x_{2i} + L(\mathbf{x}),$$

where L is linear and $2h$ is the rank of B. It is easily seen that a further affine transformation changes this representation to

$$\sum_{i=1}^{h} x_{2i-1}x_{2i} + L',$$

where $L' = 0$, 1, or x_{2h+1}.

(12.8) Lemma. *The number of points $(x_1,\ldots,x_{2h}) \in \mathbb{F}_2^{2h}$ for which $\sum_{i=1}^{h} x_{2i-1}x_{2i} = 0$ is $2^{2h-1}+2^{h-1}$.*

PROOF. If $x_1 = x_3 = \ldots = x_{2h-1} = 0$, then there are 2^h choices for $(x_2,x_4,\ldots,x_{2h})$, and if not all of them are 0, then there are 2^{h-1} choices. So the number of zeros is $2^h + (2^h-1)2^{h-1}$. □

The following lemma is a direct consequence of the representation of a quadratic form given above and Lemma 12.8.

(12.9) Lemma. *Let m be even. If $Q(\mathbf{x})$ is a quadratic form corresponding to a symplectic form B of rank m, then the coset of $\mathcal{R}(1,m)$ determined by $Q(\mathbf{x})$ has 2^m words of weight $2^{m-1}-2^{m/2-1}$ and 2^m words of weight $2^{m-1}+2^{m/2-1}$.* □

Quadratic forms of smaller rank would lead to smaller weights. For suppose that a symplectic form B has even rank $r < m$, and let W be the radical of B (of dimension $m-r$), and U a complement for W. Then there is a quadratic form Q polarizing to B which has weight $2^{r-1}-2^{r/2-1}$ on U and vanishes identically on W; then Q has weight $|W|\cdot(2^{r-1}-2^{r/2-1}) = 2^{m-1}-2^{m-r/2-1}$ overall.

We now wish to form a code C by taking the union of cosets of $\mathcal{R}(1,m)$ corresponding to certain quadratic forms $Q_1, Q_2, \ldots, Q_l$ (or the associated symplectic forms $B_1, B_2, \ldots, B_l$). So $\mathcal{R}(1,m) \subset C \subset \mathcal{R}(2,m)$. Since we are interested in the minimum distance of C, we must consider the minimum weights of the cosets defined by the forms $Q_i - Q_j$ $(i \neq j)$. Of course the best thing that we can achieve is that each difference $Q_i - Q_j$ corresponds to a symplectic form of maximal rank, that is, a *nonsingular* symplectic form. How large can l be? Since the symplectic forms correspond to skew-symmetric matrices with zero diagonal, we have $l \le 2^{m-1}$ (no two matrices may have the same first row).

(12.10) DEFINITION. Let m be even. A set of 2^{m-1} symplectic matrices of size m such that the difference of any two distinct elements is nonsingular is called a *Kerdock set.*

We shall show that it is indeed possible to construct Kerdock sets. Therefore the following definition makes sense.

(12.11) DEFINITION. Let m be even $(m \ge 4)$. Let $l = 2^{m-1}$ and let $Q_1, \ldots, Q_l$ be a Kerdock set. The nonlinear code $\mathcal{K}(m)$ of length $n = 2^m$ consisting of the cosets of $\mathcal{R}(1,m)$ corresponding to the forms Q_i $(1 \le i \le l)$ is called a *Kerdock code.*

Observe that $|\mathcal{K}(m)| = l|\mathcal{R}(1,m)| = 2^{2m}$ and that by Lemma 12.9 the minimum distance of $\mathcal{K}(m)$ is $2^{m-1} - 2^{m/2-1}$. As we saw above, a subcode of $\mathcal{R}(2,m)$ containing $\mathcal{R}(1,m)$ cannot have a larger minimum distance and with this distance it cannot have more words than $\mathcal{K}(m)$.

Remark that in the case $m = 4$ we find a code of length 16 with 2^8 words and $d = 6$. This is the (unique) Nordstrom–Robinson code treated in Chapter 11, Exercise 2.

For a nonlinear code C there is an analog of the weight enumerator, the so-called *distance enumerator* with coefficients A_i, where

$$A_i := |C|^{-1}|\{(\mathbf{x},\mathbf{y}) : \mathbf{x} \in C, \mathbf{y} \in C, d(\mathbf{x},\mathbf{y}) = i\}|.$$

From Lemma 12.9 we find the distance enumerator of $\mathcal{K}(m)$. We have

$$A_0 = 1,\ A_{2^{m-1} \pm 2^{m/2-1}} = 2^m(2^{m-1} - 1),\ A_{2^{m-1}} = 2^{m+1} - 2,\ A_{2^m} = 1.$$

If we substitute the distance enumerator in MacWilliams' relation (Theorem (9.14)) we actually find a polynomial with integer coefficients B_i (for $A^\perp$). This is in fact the distance enumerator of an extended Preparata code (see Chapter 16). There is no explanation for this strange fact!

The facts that $B_1 = \ldots = B_5 = 0$ and that $A(z)$ has only four coefficients A_i with $i > 0$, $A_i \neq 0$ have the following interesting consequence. This depends on a theorem of Delsarte (1973); we defer the proof until Chapter 14 (see (14.26)).

(12.12) Proposition. *The words of any fixed weight in $\mathcal{K}(m)$ form a 3-design.* □

Constructions of Kerdock sets.

We sketch a construction of Kerdock sets due to Dillon (1974), Dye (1977), and Kantor (1983). We shall describe the easiest construction of a Kerdock set, working out the case of four by four matrices in detail.

We consider a vector space V of dimension $4n$ over $\mathbb{F}_2$ with the quadratic form

$$Q(\mathbf{x}) := \sum_{i=1}^{2n} x_i x_{2n+i}.$$

This is called a space of type $\Omega^+(4n, 2)$. The set of singular vectors $\mathbf{x}$ (that is, those with $Q(\mathbf{x}) = 0$) is a *quadric* which we denote by $\hat{Q}$. A subspace of V is called *totally isotropic* if $\langle \mathbf{x}, \mathbf{y} \rangle = 0$ for any two points $\mathbf{x}$, $\mathbf{y}$ in the subspace. If $Q(\mathbf{x}) = 0$ for every $\mathbf{x}$ in a subspace, then we call the subspace *totally singular.* Clearly a totally singular subspace is also totally isotropic, but not conversely. The quadric contains $2n$-dimensional subspaces (maximal totally singular subspaces of V). These can be divided into two families (types). Two totally singular subspaces are of the same type if their intersection has even dimension. Each totally singular $2n-1$-space is contained in exactly two totally singular $2n$-spaces, one of each type (see Fig. 12.1). The symplectic form $B(\mathbf{x}, \mathbf{y})$ associated with Q is now denoted by $\langle \mathbf{x}, \mathbf{y} \rangle$. If $\mathbf{x}$ is any vector, then $\mathbf{x}^\perp$ is the hyperplane $\{\mathbf{y} : \langle \mathbf{x}, \mathbf{y} \rangle = 0\}$. It contains $\mathbf{x}$.

Consider a vector $\mathbf{x}$ not on $\hat{Q}$. Then we have a natural map from $\mathbf{x}^\perp$ to the symplectic $(4n-2)$-dimensional space $W := \mathbf{x}^\perp/\mathbf{x}$. Note that $\mathbf{x}^\perp$ meets $\hat{Q}$ in (several) subspaces of dimension $2n - 1$ (see Fig. 12.1).

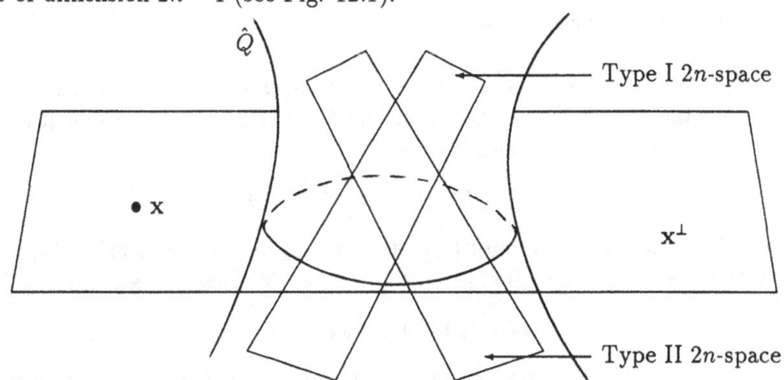

Fig. 12.1. A quadric

We now introduce two types of objects, both called *spreads.*

(12.13) Definition. A *spread* on $\hat{Q}$ (an *orthogonal* spread) is a set Σ of $2^{2n-1} + 1$

totally singular $2n$-spaces such that each nonzero vector of $\hat{Q}$ is in exactly one of them.

(12.14) DEFINITION. A spread on a symplectic space W of dimension $4n-2$ (a *symplectic spread*) is a set Σ' of $2^{2n-1}+1$ totally isotropic subspaces of dimension $2n-1$ such that each nonzero vector of W is in exactly one of them.

The map from $\mathbf{x}^{\perp}$ to W bijectively maps totally singular subspaces to totally isotropic subspaces. A spread Σ on $\hat{Q}$ leads to a spread Σ' on W. Conversely, given a space W and a spread Σ', we identify W with $\mathbf{x}^{\perp}/\mathbf{x}$ and then from Σ' we find Σ by requiring that all elements of Σ are of type I.

EXAMPLE. Take $V := \Omega^{+}(8,2)$, $Q(\mathbf{x}) := x_1x_5 + x_2x_6 + x_3x_7 + x_4x_8$. Here

$$\langle \mathbf{x}, \mathbf{y} \rangle = (x_1y_5 + x_5y_1) + \ldots + (x_4y_8 + x_8y_4).$$

As we saw in Lemma 12.8, there are 135 nonzero singular vectors. A spread on $\hat{Q}$ consists of nine totally singular 4-spaces ($135 = 9 \cdot 15$). For the point $\mathbf{x} \notin \hat{Q}$ we take (1,0,0,0,1,0,0,0). Then $\mathbf{x}^{\perp} = \{\mathbf{y} \in V : y_1 = y_5\}$ and the space $W = \mathbf{x}^{\perp}/\mathbf{x}$ is the set of vectors $(x_2, x_3, x_4, x_6, x_7, x_8)$ with as associated form $x_2x_6 + x_3x_7 + x_4x_8$.

We consider a spread Σ on $\hat{Q}$ containing the two spaces

$$E := \{(x_1, x_2, x_3, x_4, 0, 0, 0, 0) : x_i \in \mathbb{F}_2\}$$
$$F := \{(0, 0, 0, 0, x_5, x_6, x_7, x_8) : x_i \in \mathbb{F}_2\}.$$

The mapping from Σ to Σ' produces the spread with

$$E' := \{(x_2, x_3, x_4, 0, 0, 0)\} \text{ and } F' := \{(0, 0, 0, x_6, x_7, x_8)\}$$

and seven other 3-spaces on W. □

We use the example to illustrate our main point. We claim that it is easy to construct the spread Σ' directly. Consider $\mathbb{F}_8$ with a primitive element α satisfying $\alpha^3 + \alpha + 1 = 0$. For this choice we have

$$\mathrm{Tr}(\xi + \eta\alpha + \zeta\alpha^2) = \xi,$$

where Tr is the trace function : $\mathbb{F}_8 \mapsto \mathbb{F}_2$. Let W_0 be the 2-dimensional space over $\mathbb{F}_8$ with the symplectic form $f(\mathbf{x}, \mathbf{y}) := x_1y_2 + x_2y_1$. We describe a point of W_0 by

$$(x_2 + x_3\alpha + x_4\alpha^2, x_6 + x_8\alpha + x_7\alpha^2). \qquad (12.14)$$

We find the symplectic form $\mathrm{Tr}\, f$ on the space $W = Sp(6,2)$, obtained by considering W_0 as a space over $\mathbb{F}_2$. We have chosen coordinates in such a way that $\mathrm{Tr}\, f$ corresponds to the quadratic form $x_2x_6 + x_3x_7 + x_4x_8$. From now on we write the vectors of W as $(., x_2, x_3, x_4, ., x_6, x_7, x_8)$ to make the identification of W and $\mathbf{x}^{\perp}/\mathbf{x}$ more clear. A spread Σ' is obtained by simply taking all 1-subspaces of W_0, that is, the sets

$$\{\lambda(1,0) : \lambda \in \mathbb{F}_8\}, \quad \{\lambda(\xi, 1) : \lambda \in \mathbb{F}_8\}, (\xi \in \mathbb{F}_8),$$

and considering them over $\mathbb{F}_2$, producing nine totally isotropic subspaces of W which contain every nonzero point once.

The identification of W with $\mathbf{x}^\perp/\mathbf{x}$ produces two totally singular subspaces on $\hat{Q}$ (one of each type) for each element of Σ'. We give a few examples (here we list the elements of the sets in Σ', respectively Σ; each x_i runs through $\mathbb{F}_2$):

$$(1)\quad \lambda(1,0) \mapsto (.,x_2,x_3,x_4,.,0,0,0) \in \Sigma' \mapsto \left\{ \begin{matrix} (x_1,x_2,x_3,x_4,0,0,0,0) \\ (0,x_2,x_3,x_4,x_5,0,0,0) \end{matrix} \right.$$

$$(2)\quad \lambda(0,1) \mapsto (.,0,0,0,.,x_6,x_7,x_8) \in \Sigma' \mapsto \left\{ \begin{matrix} (0,0,0,0,x_5,x_6,x_7,x_8) \\ (x_1,0,0,0,0,x_6,x_7,x_8) \end{matrix} \right.$$

where in each case the upper subspace is of type I, the lower of type II. At this point we have reproduced the elements E and F of Σ. We leave the image of the line $\lambda(1,1)$ as an exercise (Exercise 8).

Now we shall show how to obtain Kerdock sets from spreads. We have chosen E and F such that $E \oplus F = V$ and such that their basis vectors correspond to our representation of Q. On V we consider linear transformations

$$\mathbf{x} \mapsto \mathbf{x}\begin{pmatrix} I & O \\ M & I \end{pmatrix}.$$

By straightforward substitution we see that these transformations preserve Q (that is, they fix $\hat{Q}$ as a set) iff $M^\top = -M$ and M has zero diagonal, that is, M is a symplectic matrix of size 4. Clearly the mapping $M \mapsto M^* := \left(\begin{smallmatrix} I & O \\ M & I \end{smallmatrix}\right)$ is an isomorphism of the group P of symplectic matrices to the group P^* of matrices $\left(\begin{smallmatrix} I & O \\ M & I \end{smallmatrix}\right)$ fixing Q.

We claim that there is a 1-1 correspondence between Kerdock sets in P and spreads in $\hat{Q}$. Given a Kerdock set $\mathcal{K}$ in P, consider the corresponding set $\mathcal{K}^*$ in P^* and form

$$\mathcal{S}(\mathcal{K}) := \{E\} \cup \{F^g : g \in \mathcal{K}^*\}.$$

If $g_1 = \left(\begin{smallmatrix} I & O \\ M_1 & I \end{smallmatrix}\right)$, $g_2 = \left(\begin{smallmatrix} I & O \\ M_2 & I \end{smallmatrix}\right)$, then $F^{g_1} \cap F^{g_2} = \{\mathbf{0}\}$ iff $M_1 - M_2$ is nonsingular. Hence $\mathcal{S}(\mathcal{K})$ is a spread on $\hat{Q}$. Conversely, consider a spread Σ containing E and F (as in our example). Define

$$K(\Sigma) := \{M \in P : FM^* \in \Sigma \backslash \{E\}\}.$$

Since any two members of Σ intersect only in $\mathbf{0}$ and $|K(\Sigma)| = 2^{2n-1}$, we see that $K(\Sigma)$ is a Kerdock set.

Example. We continue the example started above. Consider the element $F_1 := \{(a,b,c,d,b,a,d,c) : a,b,c,d \in \mathbb{F}_2\}$ of the spread Σ. We solve $F\left(\begin{smallmatrix} I & O \\ M & I \end{smallmatrix}\right) = F_1$ and find

$$M = \begin{pmatrix} 0 & 1 & 0 & 0 \\ 1 & 0 & 0 & 0 \\ 0 & 0 & 0 & 1 \\ 0 & 0 & 1 & 0 \end{pmatrix}.$$

Therefore we have to take the form $x_1x_2 + x_3x_4$ as an element of $\mathcal{K} = K(\Sigma)$. □

Much more on this construction, etc. can be found in Kantor (1983), including a relation between spreads, Kerdock sets, and translation planes. One of the interesting consequences of Kantor's work is that there are extremely many inequivalent Kerdock codes of a given length. In Kantor (1984) the following is proved.

(12.15) Proposition. *If $m-1$ is odd and composite, then there are more than $2^{\frac{1}{2}\sqrt{m}}$ pairwise inequivalent Kerdock codes $\mathcal{K}(m)$ of length 2^m.*

Kerdock sets have been used for the construction of various combinatorial objects. We discuss two of these.

(12.16) EXAMPLE. The first examples form a class of proper partial geometries.

Let Σ be a spread on the quadric Q^+ of type $\Omega(4n,2)$. Now we define an incidence structure $\mathcal{G}(n)$ as follows: the points are the non-singular vectors $\mathbf{x}$ (those not on the quadric); the lines are the $(2n-1)$-spaces which are contained in a member of Σ; and the point $\mathbf{x}$ and line H are incident if $H \subseteq \mathbf{x}^\perp$.

Let H be a line. Then $H^\perp$ has dimension 2n+1 and contains H; so it contains three $2n$-spaces containing H. But H lies in two $2n$-spaces on the quadric Q^+, and so in just one more $2n$-space K. Then the points of the incidence structure incident with H are just the 2^{2n-1} non-singular vectors of $K \setminus H$. Moreover, we see that, if two points $\mathbf{x}, \mathbf{y}$ are incident with a line H, then $\mathbf{x}+\mathbf{y}$ is a singular vector; every line incident with $\mathbf{x}$ and $\mathbf{y}$ contains this vector, from which we see that there is only one such line.

We also see from this argument that two points are collinear if and only if their sum is in Q^+. This collinearity relation defines a strongly regular graph on the points outside Q^+, and the lines form a family of cliques in this graph, with the property that each edge lies in a unique clique. It follows that the incidence structure is a partial geometry. Its parameters are

$$s = 2^{2n-1} - 1, \quad t = 2^{2n-1}, \quad \alpha = 2^{2n-2}.$$

(12.17) EXAMPLE. Next we turn to systems of linked square designs, a generalization of square designs considered by Cameron and Seidel (1973). A *system of linked square designs* is a structure consisting of sets $X_1, \ldots, X_l$, with an incidence relation I_{ij} between X_i and X_j for each i and j, such that the following conditions hold:

(a) for each pair i, j, (X_i, X_j, I_{ij}) is a non-trivial square 2-design;
(b) for each triple i, j, k, and each choice of $x \in X_i$ and $y \in X_j$, the number of elements of X_k which are incident with both x and y depends only on i, j, k and whether or not x and y are incident.

These structures were introduced to study doubly transitive permutation groups. It is not known whether the parameters of the square designs in a linked system must necessarily all be the same or complementary. The only known examples arise from the following construction.

As we saw in Chapter 5, to each alternating bilinear form B on $V(2n, 2)$, there is a set $\mathcal{Q}(B)$ of 2^{2n} quadratic forms Q associated with B; that is, such that

$$Q(\mathbf{x}+\mathbf{y}) = Q(\mathbf{x}) + Q(\mathbf{y}) + B(\mathbf{x}, \mathbf{y}).$$

These are all the functions with zero constant term lying in a coset of $\mathcal{R}(1, 2n)$ in $\mathcal{R}(2, 2n)$. Recall that a non-singular quadratic form has type ϵ if it has $2^{2n-1} + \epsilon 2^{n-1}$ zeros, for $\epsilon = \pm$. If B_1 and B_2 are alternating bilinear forms for which $B_1 - B_2$ is non-singular, we can define an incidence relation between the sets $\mathcal{Q}(B_1)$ and $\mathcal{Q}(B_2)$, by declaring Q_1 and Q_2 incident if $Q_1 - Q_2$ has type $+$; this gives a square 2-$(2^{2n}, 2^{2n-1} + 2^{n-1}, 2^{2n-2} + 2^{n-1})$ design. Now it is possible to show the following:

(12.18) Proposition. *Let B_i ($i = 1, 2, 3$) be alternating bilinear forms on $V(2n, 2)$ such that the difference of any pair is non-singular. Let $Q_i \in \mathcal{Q}(B_i)$ for $i = 1, 2$. Then the number of forms $Q_3 \in \mathcal{Q}(B_3)$ such that $Q_1 - Q_3$ and $Q_2 - Q_3$ both have type $+$, is either $2^{n-2}(2^n + 3)$ or $2^{n-2}(2^n + 1)$, depending on whether $Q_1 - Q_2$ has type $+$ or $-$.* □

Now let $B_1, \ldots, B_l$ be alternating bilinear forms, all of whose differences are non-singular. If $X_i = \mathcal{Q}(B_i)$ for $i = 1, \ldots, l$, then the incidence relations defined above give $(X_1, \ldots, X_l)$ the structure of a system of linked square designs.

In particular, we get the largest possible system from this construction (viz. $l = 2^{2n-1}$) by using a Kerdock set. Noda (1974) has shown that, for systems where all square designs have the parameters $(2^{2n}, 2^{2n-1} + 2^{n-1}, 2^{2n-2} + 2^{n-1})$, the number l of sets X_i cannot exceed 2^{2n-1}.

The RM codes are examples of a larger class of codes known as *Euclidean Geometry Codes* (see e.g. J.-M. Goethals (1973)).

Closely related to these codes are the *Projective Geometry Codes.* We give only one example.

(12.19) DEFINITION. Let p be a prime, $q = p^\alpha$, and let A be the incidence matrix of points and hyperplanes of $PG(m, q)$. The rows of A generate a code $C^\perp$ over the alphabet $\mathbf{F}_p$. The dual code C is called a *projective geometry code.*

It can be shown (see Goethals (1973)) that $C^\perp$ has dimension $1 + \binom{m+p-1}{p-1}^\alpha$. An example connected to the topic of our next chapter is the case $p = \alpha = m = 2$. In this case A is the incidence matrix of $PG(2, 4)$. The code $C^\perp$ has dimension 10. By straightforward calculation one can show that $C^\perp$ has exactly 21 words of weight 5. This implies that a codeword has weight 5 iff it is a line of $PG(2, 4)$. So A can be retrieved from its linear span. This is another example of a code for which the vectors of minimum weight ($\neq 0$) form a 2-design.

We give one more example linking RM codes to combinatorial theory. Consider the first order RM code of length $n = 2^m$. We have already observed that the codewords come in pairs (parallel hyperplanes) with distance n. From each pair we choose one word. Since the code has dimension $m+1$, we find n words. With these words we form a square matrix (n by n). Then replace all 0s by -1. Because any two of the chosen words have distance $\frac{1}{2}n$, the resulting matrix is a *Hadamard matrix*. In fact, it is a Sylvester matrix (see Chapter 1, and especially Exercise 2).

We briefly discuss *equidistant codes* and a connection to geometry. These codes have the property that any two distinct codewords have the same distance. The code treated at the end of Chapter 9 was an example with distance 6. The rows of a Hadamard matrix, with -1s replaced by 0s, also form an equidistant code. How large can an equidistant code be?

If the distance d is odd, then the code obviously can have only two codewords. If $d = 2k$ and the code has m words of length n (n sufficiently large), then it is not difficult to show that

$$m \leq k^2 + k + 2$$

unless the code is trivial. (Here, a code is called *trivial* if the m by n matrix with the codewords as rows has the property that every column has $m-1$ or m equal entries (see Deza (1973))).

Now, let A be the incidence matrix of a projective plane of order k. To A we adjoin $k-1$ columns of ones and then we adjoin a row of zeros. The resulting matrix has as its rows the codewords of an equidistant code with k^2+k+2 words with mutual distance $d = 2k$. It is much more difficult to show that the converse is true, namely that if an equidistant code with k^2+k+2 words and mutual distance $2k$ exists, then also a projective plane of order k exists (cf. Van Lint (1969)). We remark that it has been shown that in the case that $k = 6$, that is, $d = 12$ (where no projective plane exists), the maximal number of words is 32 (cf. J. I. Hall *et al.* (1977)).

Exercises

1. Prove that in the expansion of $(x+1)^n$ in $\mathbb{F}_2[x]$, the coefficient of x^k is 1 iff the binary expansion of n has a 1 in every position where the binary expansion of k has a 1. Let $n = 2^m$. Let

$$(x+1)^k = \sum_{i=1}^{n-1} a_i x^{n-1-i}.$$

Show that $\mathbf{a} = (a_0, \ldots, a_{n-1}) \in \mathcal{R}(r, m)$ iff $w(k) \geq m - r$.

2. Let $n = 2^m$. Show that for any $\mathbf{x} \in \mathbb{F}_2^n$ there is a codeword $\mathbf{c}$ in $\mathcal{R}(1, m)$ such

that $d(\mathbf{x}, \mathbf{c}) \le \frac{1}{2}(n - \sqrt{n})$. (Hint: use the relation between $\mathcal{R}(1, m)$ and Hadamard matrices.)

3. Let m be even. Show that there is a vector $\mathbf{x}$ in $\mathbb{F}_2^m$ such that for all $\mathbf{c} \in \mathcal{R}(1, m)$ one has $d(\mathbf{x}, \mathbf{c}) \ge \frac{1}{2}(n - \sqrt{n})$.

4. Suppose the code $\mathcal{R}(2, 5)$ is used and the received message is

$$(10110100101101001011000000001111).$$

Decode this message. (This is tedious! Useful to do it once and to realize that this is work for computers.)

5. Delete the top row and the first column of the generator matrix of $\mathcal{R}(1, 5)$. We call the new matrix G. The matrix G is the generator matrix of a linear code C. Show that all the words $\ne \mathbf{0}$ in C have weight 16. Find seven columns of G such that every word in C has either seven 0's or four 1's on these seven positions. Use this idea to construct a [21,5,10] code.

6. Let $\mathbf{w}$ be a word of the extended binary Golay code $\mathcal{G}$, with weight 8 and support A. Prove that

(a) the set of words with support disjoint from A form the Reed–Muller code $\mathcal{R}(1, 4)$;

(b) the restrictions to the complement of A of all words in $\mathcal{G}$ form the Reed–Muller code $\mathcal{R}(2, 4)$;

(c) the restrictions to the complement of A of words whose support meets A either in the empty set, or in a 2-set containing a fixed $a \in A$, form the Kerdock code $\mathcal{K}(4)$.

Hence give an alternative proof of (12.12) for $\mathcal{K}(4)$.

Show that the restrictions to the complement of A of words whose supports meet A in $\emptyset$, $\{a, b\}$, $\{a, c\}$ or $\{b, c\}$ (for fixed $a, b, c \in A$) form a $[16, 7, 6]$ code. Can you recognize this code?

Remark. From (b), we see that there is a bijection between the set of symplectic forms on $\mathbb{F}_2^4$ and the partitions of an 8-set A into (at most) two parts of even cardinality. The zero form corresponds to the partition $\{\emptyset, A\}$; non-zero singular forms to partitions into parts of size 4; and non-singular forms to partitions into parts of sizes 2 and 6.

7. Let $m = 2k$. Consider the vectors in $\mathbb{F}_{2^m} \backslash \{0\}$ on the quadric with equation $x_1x_2 + \ldots + x_{2k-1}x_{2k} = 0$. We take these as columns of a matrix G. The matrix G generates a linear code C. Determine the weight enumerator of C.

8. Consider the example given in the section on Kerdock sets. The elements E and F of Σ were constructed. Show that the line $\lambda(1, 1)$ leads to the totally singular

4-spaces with points (a,b,c,d,d,a,b,c) and (a,a,c,d,b,b,d,c). Which one belongs to Σ?

9. Consider the 1-space $\lambda(\alpha,1)$ in W_0 and show that by (12.14) it corresponds to the totally singular 4-space F_α of type I on $\hat{Q}$ with basis vectors (0,0,1,0,0,1,0,0), (1,0,0,1,1,0,0 ,1), (1,1,1,0,1,0,1,0), and (1,1,0,1,0,0,1,0). Then solve $F\begin{pmatrix} I & O \\ M & I \end{pmatrix} = F_\alpha$ to find the element $x_1x_3+x_1x_4+x_2x_3+x_3x_4$ in $\mathcal{K}$. Check that this form and $x_1x_2+x_3x_4$ have nonsingular difference.

10. Represent a quadratic form in x_1 to x_4 by a graph on four vertices 1,2,3,4 by letting a term x_ix_j correspond to the edge $\{i,j\}$. Which graphs correspond to nonsingular quadratic forms? Use these pictures to construct $\mathcal{K}(4)$.

11. Consider the words in $\mathcal{G}_{11}$ ending in five 0's. Show that if we delete these five 0's, the resulting code is the Z used in the construction of the partial geometry with $s=t=5$, $\alpha=2$, in Example (7.20). So there is a natural bijection between the cosets of $\mathcal{G}_{11}$ in F_3^{11} and those of Z in F_3^6. Show that the tripartite incidence graph Γ of the three 'linked' partial geometries is a subgraph of the graph of Proposition (11.12).

12. Prove that $\mathrm{Aut}(\mathcal{R}(r,m)) = \mathrm{AGL}(m,2)$ if $0 < r < m-1$.

13. Show that the coset of $\mathcal{R}(1,2n)$ containing the function

$$x_1x_{n+1} + x_2x_{n+2} + \ldots + x_nx_{2n} + x_1x_2\ldots x_i$$

for $2 \le i \le n$, contains words of only two different weights (and has the weight distribution given by (12.9)).

Deduce that the supports of the words of weight $2^{2n-1}+2^{n-1}$ form a square 2-design having the *symmetric difference property* of Chapter 5, viz., the symmetric difference of any three blocks is either a block or the complement of one.

Show that these designs, for different values of i, are pairwise non-isomorphic. (HINT: The hyperplanes of AG$(2n,2)$ can be obtained as symmetric differences of pairs of blocks. Prove that, no matter how the affine space is coordinatized, the degree of a polynomial function is invariant.)

REMARK. Only in the quadratic case ($i=2$) does the design admit the translation group of the affine space as a group of automorphisms.

The examples are due to N. J. Patterson.

14. Prove that the support of any word of minimum weight in $\mathcal{R}(r,m)$ is an $(m-r)$-flat. [HINT: Show that a hyperplane meets such a support set in 0 or at least $\frac{1}{2}|S|$ points, and hence in 0, $\frac{1}{2}|S|$, or $|S|$ points. Show that $|S\cap H| = \frac{1}{2}|S|$ for some hyperplane H; using induction, deduce that $S\cap H$ is an $(m-r-1)$-flat, and that S is the union of two translates of this flat.]

13. Self-orthogonal codes and projective planes

We shall first look at some self-dual codes. A first example was given in Chapter 9, namely the [8,4] extended Hamming code over $\mathbb{F}_2$. Actually, this is an example of a larger class, that follows from Theorem (12.3). The code $\mathcal{R}(r, 2r+1)$ is self-dual. We have seen some designs related to these codes and in a later chapter it will be shown that quite often self-duality is a useful property to have when trying to construct designs from codes.

Now we shall consider a special class of ternary self-dual codes. The codes are cyclic with length $n \equiv -1 \pmod 3$, where n has the following property. If α is a primitive n-th root of unity, then for $0 < i < n$ the powers α^i and α^{-i} are not zeros of the same irreducible polynomial. Now choose a polynomial $g(x)$ that divides $x^n - 1$, that has $g(1) \neq 0$, and such that for every i with $0 < i < n$ either $g(\alpha^i) = 0$ or $g(\alpha^{-i}) = 0$. Let C be the cyclic (ternary) code with generator $g(x)$. Clearly C has dimension $\frac{1}{2}(n+1)$ and since $C^{\perp}$ has generator $(x-1)g(x)$, the code $C^{\perp}$ is the subcode of C consisting of the words $c(x) \in C$ with $c(1) = 0$. It follows that C has a basis consisting of $\mathbf{1}$ and a set of basis vectors of $C^{\perp}$. Using the fact that $n+1 \equiv 0 \pmod 3$ it then follows that $\overline{C}$ is self-dual.

(13.1) EXAMPLE. Take $n = 11$ and let C be the code with generator $m_1(x)$. So the zeros are α^i with $i = 1, 3, 4, 5, 9$. By Theorem (10.6), the minimum distance of C is at least 4. Since $\overline{C}$ is self-dual, it must have minimum weight at least 6 (and hence C in fact has $d = 5$). So $\overline{C}$ is a ternary self-dual [12,6,6] code, necessarily $\mathcal{G}_{12}$!

The main topic in this chapter are (self-dual) codes generated by the incidence matrix of a projective plane. We have seen an example in the section on projective geometry codes.

Let A be the incidence matrix of PG$(2, n)$. We consider the subspace C of $\mathbb{F}_2^{n^2+n+1}$ that is generated by the rows of A. If n is odd it is easy to see what the code is. If we calculate the sum of the rows of A that have a 1 in a fixed position, the result is a row with a 0 in that position and 1's elsewhere. These rows generate the subspace of $\mathbb{F}_2^{n^2+n+1}$ consisting of all words of even weight. Since C obviously has no words of odd weight, this subspace is C. The case of even n is more difficult. We are mainly interested in $n \equiv 2 \pmod 4$. (We use $\mathrm{PG}(2,n)$ for any projective plane of order n, in contrast to our earlier notation where this symbol stood for the projective plane over $\mathbb{F}_n$.)

(13.2) Theorem. *If $n \equiv 2 \pmod 4$ then the rows of the incidence matrix of $\mathrm{PG}(2,n)$ generate a code C with dimension $\frac{1}{2}(n^2+n+2)$.*

PROOF. (i) Since n is even, the code $\overline{C}$ is self-orthogonal, that is, $\overline{C} \subset (\overline{C})^\perp$. Therefore $\dim C \le \frac{1}{2}(n^2+n+2)$.

(ii) Let $\dim C = r$ and let $k := n^2+n+1-r = \dim C^\perp$. Let H be a parity check matrix of rank k for C and assume that the coordinate places have been permuted in such a way that H has the form $(I_k\ P)$.

Define $N := \begin{pmatrix} I_k & P \\ O & I_r \end{pmatrix}$. Interpret A and N as rational matrices. Then

$$\det AN^\top = \det A = (n+1)n^{\frac{1}{2}(n^2+n)}.$$

Since all entries in the first k columns of $AN^\top$ are even, we find that $2^k \mid \det A$. It then follows that $r \ge \frac{1}{2}(n^2+n+2)$.

From (i) and (ii) the theorem follows. □

We remark that the theorem can also be proved using the invariant factors of A. That method has the advantage that it generalizes to other characteristics. It will be illustrated in the proof of Theorem (13.10). The following proposition is an immediate consequence of Theorem (13.2).

(13.3) Proposition. *If $n \equiv 2 \pmod 4$, then the rows of the incidence matrix A of $\mathrm{PG}(2,n)$ generate a code C, for which $\overline{C}$ is self-dual.* □

We continue with a few other properties of $\mathrm{PG}(2,n)$, where $n \equiv 2 \pmod 4$, formulated in terms of coding theory.

(13.4) Theorem. *The code C of Proposition (13.3) has minimum weight $n+1$ and every word of minimum weight is a line in $\mathrm{PG}(2,n)$.*

PROOF. Let $\mathbf{v} \ne \mathbf{0}$ be a codeword with $w(\mathbf{v}) = d$. Since every line has a 1 as overall parity check, we see that:
(1) if d is odd then $\mathbf{v}$ meets every line at least once;
(2) if d is even then every line through a fixed point of $\mathbf{v}$ meets $\mathbf{v}$ in a second point.
In case (2) we immediately have $d > n+1$. In case (1) we find $(n+1)d \ge n^2+n+1$, that is, $d \ge n+1$. If $w(\mathbf{v}) = n+1$ then there is a line l of $\mathrm{PG}(2,n)$ that meets $\mathbf{v}$

in at least three points. If there is a point of l not on $\mathbf{v}$, then every line $\neq l$ through this point meets $\mathbf{v}$ (by (1)). This would yield $d \geq n+3$. □

We remind the reader that 'type II ovals' were defined in Chapter 1.

(13.5) Theorem. *The words of weight $n+2$ in C are precisely the type II ovals of* PG$(2,n)$.

PROOF. (i) Let $\mathbf{v} \in C$ and $w(\mathbf{v}) = n+2$. Every line meets $\mathbf{v}$ in an even number of points. Let l be a line and suppose that $\mathbf{v}$ and l have $2a$ points in common. Each of the n lines $\neq l$ through one of these $2a$ points meets $\mathbf{v}$ at least once more. Therefore $2a + n \leq n+2$, that is, $a = 0$ or $a = 1$.

(ii) let V be a type II oval. Let S be the set of $\frac{1}{2}(n+1)(n+2)$ distinct lines of PG$(2,n)$ through the pairs of points of V. Each point not in $\mathbf{v}$ is on $\frac{1}{2}(n+2)$ of these lines; each point of $\mathbf{v}$ is on $n+1$ lines of S. Since $n \equiv 2 \pmod 4$, it follows that $\mathbf{v}$ is the sum of the lines of S, that is, $\mathbf{v} \in C$. □

We do not treat the case $n \equiv 0 \pmod 4$ in general but we give one example because it plays a rôle in our treatment of the binary Golay code. (See Chapter 1.)

(13.6) EXAMPLE. Let S be a set of three points a_1, a_2, a_3 in PG(2,4) not on one line. There are nine lines in PG(2,4) that contain exactly one of these three points. These form the set $\mathcal{L}$. Each point on one of the lines through two points of S is on exactly one line in $\mathcal{L}$. So, as elements of $\mathbb{F}_2^{21}$ these lines are independent. (We are in fact looking at points of a unital and the nine tangents.) Observe that the line through a_1 and a_2 is independent of the previous nine lines. This implies that the code C generated by the incidence matrix A of PG(2,4) has dimension at least 10. Clearly $\overline{C}$ is again self-orthogonal and therefore C has dimension at most 11. The even-weight subcode of C is generated by the 20 words obtained by adding the top row of A to the other rows. These rows all have weight 8 and they are mutually orthogonal. This means that the even-weight subcode of C is doubly even. Now we are in a different situation than in Theorem (13.5). The ovals of PG(2,4) correspond to words of weight 6 in $C^\perp$ but they are apparently not in C. Hence C does not have dimension 11 but dimension 10. The even-weight subcode of C (with dimension 9) is also a subcode of $C^\perp$. This code and one orbit of ovals generate an even-weight code of dimension 10; adjoining another orbit of ovals produces an even-weight code of dimension 11 containing all the ovals. This is $C^\perp$.

For a self-dual $[n,k]$ code C over $\mathbb{F}_q$ we find from Theorem (9.14) the following relation for the weight enumerator $A(\xi,\eta)$:

$$A(\xi,\eta) = q^{-k} A(\eta - \xi, \eta + (q-1)\xi), \tag{13.7}$$

where $k = \frac{1}{2}n$. This means that the polynomial $A(\xi,\eta)$ is invariant under the linear transformation with matrix $q^{-\frac{1}{2}}\begin{pmatrix} -1 & q-1 \\ 1 & 1 \end{pmatrix}$. If $q = 2$, all codewords of C have even weight, that is, $A(\xi,\eta)$ is invariant under the transformation $\xi \mapsto -\xi$. The two transformations generate the dihedral group $\mathcal{D}_8$. It was shown by A. M. Gleason (see Berlekamp *et al* (1972), Gleason (1978), and Mallows and Sloane (1973)) that the ring of polynomials in ξ and η, that are invariant under this group, is the free ring generated by $\xi^2+\eta^2$ and $\xi^2\eta^2(\xi^2-\eta^2)^2$.

Much of the work described in this chapter was motivated by attempts to construct a projective plane of order 10 or to prove its nonexistence (and this finally led to the latter of the two). We start by sketching the first significant result in the long history of these attacks.

Suppose a plane of order 10 exists. From (13.3) to (13.5) we know that the incidence matrix of this plane generates a code C for which $\overline{C}$ is a [112,56] self-dual code and that the weight enumerator $A(\xi,\eta)$ of C has coefficients

$$A_0 = 1, \quad A_1 = A_2 = \ldots = A_{10} = 0, \quad A_{11} = 111.$$

In the same way as before, we see that $\overline{C}$ is doubly even, so $A_{13} = A_{14} = 0$. By using (13.7) or Gleason's result, one then sees that $A(\xi,\eta)$ is uniquely determined if we know A_{12}, A_{15}, and A_{16}. MacWilliams *et al* (1973) investigated the codewords of weight 15 in C. It was assumed that the incidence matrix of the plane led to a code C for which $A_{15} \neq 0$. Geometric arguments of the same type as the ones we used to prove Theorems (13.4) and (13.5) severely restricted the possible corresponding geometric configurations and this resulted in a particular configuration of 15 lines that had to be part of the plane. In other words, a part of the incidence matrix was known. A computer search then showed that starting with this part it was not possible to complete the matrix to an incidence matrix of a plane of order 10. At that point it was therefore known that if such a plane exists, then $A_{15} = 0$. Later Bruen and Fisher (1973) pointed out that this same result also follows from a computer result of R. H. F. Denniston concerning the non-existence of 6-arcs not contained in 7-arcs in a plane of order 10 (see Exercise 5).

Subsequent attacks on the plane went along the same lines. Coefficients of the weight enumerator were related to special geometric subconfigurations of the plane and these were used as input for computer searches. Several years of computing finally led to the nonexistence of the plane.

As a historically interesting remark we include an attempt to construct the plane with the methods of this chapter. It is based on Theorem (13.4). Let D be the incidence matrix of a 2-(56,11,2) design. Four of these designs are known. The matrix $G := (I_{56}\ D)$ generates a binary [112,56] code C. The code is doubly even and self-dual. We discuss only the case where the 2-(56,11,2) design corresponds to the Gewirtz graph. In that case D is symmetric and therefore $(D\ I_{56})$ is also a generator matrix for C. It follows that in order to show that C has minimum weight 12, it is sufficient to consider the sum of $i \leq 6$ rows of G. Let x_j $(1 \leq j \leq 56)$ be the number of ones in the j-th column and in some given i-tuple of rows of D. Then we have

$$\sum_{j=1}^{56} x_j = 11i, \quad \sum_{j=1}^{56} \binom{x_j}{2} = 2\binom{i}{2}.$$

Therefore $\sum_{j=1}^{56} x_j(x_j - 2) = 2i^2 - 13i$, that is, at least $13i - 2i^2$ columns have exactly one 1 in these i rows. This shows that the weight of the sum of these i rows is at least $2i(7 - i)$. This is at least 12 and takes this value only if $i = 1$ or $i = 6$. Suppose that the sum of six rows of G has weight 12. This implies that $x_j = 0$ for 20 values of j, $x_j = 1$ for six values of j, and for the remaining 30 columns of D we have $x_j = 2$. So, the six rows of D (blocks of the design) have the property that no point is in more than two of them, and thus they correspond to the tangents of a (type I) oval in the 2-(56,11,2) design. If we are lucky, then the code C corresponds to the plane PG(2,10) and then we can reconstruct the plane using Theorem (13.4) (just consider the ovals of the design through a fixed point). The other designs with the same parameters can be treated in a similar way. None of the designs produced the desired result (and we now know why).

A further generalization of these ideas is based on a different inner product on $\mathbb{F}_p^{N+1}$, where $N := n^2 + n + 1$. This is the so-called *Minkowski inner product*

$$\langle \mathbf{x}, \mathbf{y} \rangle := x_1y_1 + x_2y_2 + \ldots + x_Ny_N - x_{N+1}y_{N+1}.$$

(13.8) Theorem. *Let A be the incidence matrix of a projective plane π of order n, and let C be the code generated by the rows of A over $\mathbb{F}_p$. To each row of A we adjoin the symbol -1 (so the sum of the symbols in a row is 0). The extended matrix generates a code that we call C^*. If $p\|n$, then C^* is 'self-dual' with respect to the Minkowski inner product.*

PROOF. Interpreting orthogonality and thus duality with respect to the Minkowski inner product, it is clear that $C^* \subseteq (C^*)^\perp$. The dimension of C^* is equal to the p-rank $\mathrm{rk}_p(A)$ of A (its rank regarded as a matrix over $\mathbb{F}_p$). We now appeal to the theory of invariant factors, or Smith normal form (cf. Cohn (1974), p. 279). There are integral matrices P and Q, invertible over $\mathbb{Z}$ (that is, having determinants ± 1), such that $PAQ = D$, where D is a diagonal matrix $\mathrm{diag}(d_1, d_2, \ldots, d_N)$ with $d_i \in \mathbb{Z}$, satisfying $d_i | d_{i+1}$ for $1 \le i \le N - 1$. Now $\mathrm{rk}_p(A) = \mathrm{rk}_p(D)$ is the number of invariant factors d_i which are not divisible by p. We have

$$(\det A)^2 = \det AA^\top = \det(nI + J) = (n+1)^2 n^{N-1},$$

so $\pm \det A = \det D = d_1 d_2 \cdots d_N$ is exactly divisible by $p^{(N-1)/2}$. Thus at most $\frac{1}{2}(N-1)$ of the invariant factors are divisible by p, that is, $\dim C^* \ge \frac{1}{2}(N+1)$. Since we already observed that $C^* \subseteq (C^*)^\perp$ we must have $C^* = (C^*)^\perp$. $\square$

(13.9) EXAMPLE. We give an example very similar to Example (13.1). Consider PG(2,3) in its cyclic representation corresponding to the difference set $\{0, 1, 3, 9\}$

(mod 13). The cyclic ternary code C generated by the incidence matrix A of the plane (a circulant) has generator polynomial $g(x)$, where $g(x)$ is the greatest common divisor of $x^{13}-1$ and $1+x+x^3+x^9$. This is

$$g(x) = x^6 - x^5 - x^4 - x^3 + x^2 - x + 1.$$

If α is a primitive 13-th root of unity satisfying $\alpha^3+\alpha^2-1=0$, then $g(x) = m_1(x)m_7(x)$ with zeros α^i with $i = 1,3,7,8,9,11$. By Theorem (10.6) the minimum distance of C is at least 4 and since the lines of PG(2,3) have four points, the distance is exactly 4. The dual code $C^\perp$ has generator $(x-1)g(x)$ and as in Example (13.1), it consists of the words $c(x) \in C$ with $c(1) = 0$. We take as basis for C the word **1** and six basis vectors of $C^\perp$.. Since $13+1$ is not divisible by 3, the code $\overline{C}$ is not self-dual. However with respect to the Minkowski inner product, the code C^* is self-dual. A basis is obtained by adjoining -1 to **1** and 0 to the other basis vectors. Of course in this case the fact that C^* has dimension 7 follows directly from $\deg g(x) = 6$.

The construction of Theorem (13.8) was generalized by Lander (1981), (1983). He showed the following.

(13.10) Theorem. *Let π be a projective plane of order n with incidence matrix A. Let $N := n^2+n+1$, and let p be a prime such that $p^s \| n$ $(s>0)$. Then there is a sequence*

$$\{0\} = C_{-1} \subseteq C_0 \subseteq \ldots \subseteq C_s = \mathbb{F}_p^{N+1}$$

of codes of length $N+1$ over $\mathbb{F}_p$, with the following properties:
(a) each code C_i is invariant under all automorphisms of π;
(b) C_0 is the extended code generated by the rows of A;
(c) for $-1 \le i \le s$, $C_i^\perp = C_{s-1-i}$ (relative to the Minkowski inner product);
(d) $\dim C_i$ is equal to the number of invariant factors of A not divisible by p^{i+1} for $i < s$;
(e) for $0 \le i \le s-1$, C_i has minimum weight $n+2$, and the words of weight $n+2$ in C_0 correspond to lines or possibly (if $p=2$) ovals in π. □

REMARKS. 1. If s is odd, we see from (3) that $C_{(s-1)/2}$ is a self-dual code associated with π.

2. Lander has also calculated the dimensions of the codes C_i in the case where π is Desarguesian, generalizing the known result that $\dim C_0 = 1 + \binom{p+1}{2}^s$ in this case (cf. Goethals (1973), Hamada (1973)).

3. Example (13.6) is an example of Theorem (13.10). There $p=2$, $s=2$. The matrix A has ten odd invariant factors, two are divisible by 2 but not by 4, and the others have a factor 4 (here $\det A = 5 \cdot 4^{10}$).

The methods of this chapter were used by Assmus and Van Lint (1979) to give a characterization of the biplane of order 2.

(13.11) Theorem. *Let $\mathcal{D}$ be a symmetric 2-(v, k, λ) design with $k \equiv 0 \pmod 4$, $\lambda \equiv 2 \pmod 4$, $\lambda | k$. Suppose $\mathcal{D}$ has ovals and that these ovals also form a design 2-(v, K, Λ). Then $\mathcal{D}$ is the unique 2-(7,4,2) design.*

PROOF. First we observe that the design formed by the type II ovals has $K = (k+\lambda)/\lambda$. Let A be the incidence matrix of $\mathcal{D}$ and let C be the binary code generated by the rows of A. Clearly $C \subset C^\perp$. Since v is odd we have $\dim C \le \frac{1}{2}(v-1)$. We proceed as in the proof of Theorem (13.8). Let $d_1, d_2, \dots, d_v$ be the invariant factors of A. Then $d_i | d_{i+1}$, $(1 \le i \le v-1)$, $d_1 d_2 \cdots d_v = \det A = k(k-\lambda)^{(v-1)/2}$, and $d_1 d_2 \cdots d_{v-1}$ is the g.c.d. of the minors of order $v-1$ of A. If we add all the rows of A we find a row with all entries k. It follows that $d_1 d_2 \cdots d_{v-1}$ is a divisor of $(k-\lambda)^{(v-1)/2}$. Therefore at least $\frac{1}{2}(v-1)$ of the invariant factors are odd and hence the rank of A over $\mathbb{F}_2$ is at least $\frac{1}{2}(v-1)$. Since $\mathbf{1} \in C^\perp$ but $\mathbf{1} \notin C$ (all codewords of C have even weight) we have proved that $\dim C = \frac{1}{2}(v-1)$ and $C^\perp$ is the direct sum of C and $\mathbf{1}$.

Now consider a vector of weight $d > 0$ in $C^\perp$. The same argument as in Theorem (13.5) shows that $d \ge (k+\lambda)/\lambda$, and so the ovals are vectors of minimum weight in $C^\perp$. Now, using the fact that the ovals also form a design we can repeat the argument and thus show that the minimum weight of C is k (which is achieved by the blocks of $\mathcal{D}$). Since the minimum weight of $C^\perp$ is less than k we see that each oval has the form $\mathbf{1} + \mathbf{c}$, where $\mathbf{c} \in C$. It follows that the sum of two ovals is in C. If these two ovals meet in a points, we must have $2\left(\frac{k+\lambda}{\lambda - a}\right) \ge k$. Because we know that all words in C have weight divisible by 4. we find $a = 1$ as only possibility (and hence $\lambda = 2$). This means that the design formed by the ovals of $\mathcal{D}$ is a projective plane of order $\frac{1}{2}k$ on $v = 1 + \frac{1}{2}k(k-1)$ points, and therefore $k = 4$. □

Exercises.

1. Let C be the ternary code of length 16 with generator

$$g(x) = m_1(x) m_2(x) m_4(x) m_8(x).$$

Show that C has minimum distance at least 6.

2. Consider the code generated by the incidence matrix of the projective plane of order 4. Find the number of words of weight 5,6,7,8, and 9, respectively. Give a geometric characterization of these words in each of these cases. Then determine the weight enumerator of the code.

3. Let A be the incidence matrix of PG(2,4). Suppose k rows of A sum to $\mathbf{0}$ (over $\mathbb{F}_2$). What can k be? Do the corresponding lines form a special geometric configuration?

4. The words of minimum weight in $\mathcal{R}(r, m)$ correspond to the $(m-r)$-flats (see Chapter 12, Exercise 14). Use this fact and Gleason's result on self-dual codes to find the weight enumerator of $\mathcal{R}(2,5)$.

5. Suppose that a projective plane of order 10 exists, and that **w** is a word of weight 15 in the code of this plane. Let S be the support of **w**.

Prove that any line meets S in 1, 3 or 5 points. By counting incidences between i points of S and a line, for $i = 0, 1, 2$, show that there are exactly six lines meeting S in 5 points, and that each point of S lies on two of these lines.

Hence show that this set of six lines is a 6-arc in the dual plane, which cannot be extended to a 7-arc.

6. (a) Use the MacWilliams identities to show that a doubly even self-dual code of length 16 necessarily has weight enumerator

$$\xi^{16} + 28\xi^{12}\eta^4 + 198\xi^8\eta^8 + 28\xi^4\eta^{12} + \eta^{16}.$$

(b) Find two inequivalent codes satisfying the above conditions.

14. Quadratic residue codes and the Assmus–Mattson theorem

Before starting with the actual topics of this chapter, we must prove a theorem on cyclic codes that will provide us with a very useful tool for the treatment of many codes. We use the symbol π_j to denote the permutation of positions in codewords given by $\pi_j(x^k) := x^{jk} \pmod{x^n - 1}$, where we use the identification of vectors in $\mathbb{F}_2^n$ and polynomials $\bmod(x^n - 1)$. We consider binary cyclic codes of length n, where n is odd.

(14.1) Theorem. *For every ideal V in $\mathbb{F}_2^n$, there is a unique polynomial $c(x) \in V$, called the* idempotent *of V, with the following properties:*
(a) $c(x) = c^2(x)$;
(b) $c(x)$ generates V;
(c) $\forall_{f(x) \in V}[c(x)f(x) = f(x)]$, that is, $c(x)$ is a unit-element for V;
(d) if $(j,n) = 1$, then $\pi_j(c(x))$ is the idempotent of $\pi_j V$.

PROOF. Let $g(x)$ be the generator polynomial of the ideal V and let $g(x)h(x) = x^n - 1$ in $\mathbb{F}_2[x]$. Since $x^n - 1$ has no multiple zeros, we have $(g(x), h(x)) = 1$. Therefore there are polynomials $p_1(x)$ and $p_2(x)$ such that

$$p_1(x)g(x) + p_2(x)h(x) = 1 \quad (\text{in } \mathbb{F}_2[x]). \tag{14.2}$$

We multiply both sides of this equation by $c(x)$, where $c(x) := p_1(x)g(x)$. We find

$$c^2(x) + p_1(x)p_2(x)g(x)h(x) = c(x).$$

Since in $\mathbb{F}_2^n$ (interpreted as $\mathbb{F}_2[x] \pmod{x^n - 1}$) we have $g(x)h(x) = 0$, we have proved (a).

The monic polynomial of lowest degree in the ideal generated by $c(x)$ is $(c(x), x^n - 1) = (p_1(x)g(x), g(x)h(x)) = g(x)$. This proves (b).

By (b) every $f(x)$ in V is a multiple of $c(x)$. Let $f(x) = c(x)f_1(x)$. Then $c(x)f(x) = f(x)$ by (1). This proves (c).

Since π_j is an automorphism of $\mathbb{F}_2^n$, the polynomial $\pi_j(c(x))$ is an idempotent and also a unit-element in $\pi_j V$. Since such an element is unique, (d) has been proved. □

(14.3) EXAMPLE. We consider an example in detail. Let $n = 2^m - 1$. Let $m_1(x)$ be the minimal polynomial of a primitive element α of $\mathbb{F}_{2^m}$. Then, by Theorem (10.4), $m_1(x)$ generates a Hamming code, which we now denote by H_m. Let $x^n - 1 = m_1(x)h(x)$ in $\mathbb{F}_2[x]$. We know that $h(x)$ generates a code H_m^* equivalent to $H_m^\perp$. This code H_m^* has dimension m and hence it consists of $\mathbf{0}$ and the n cyclic shifts of the word $h(x)$. (This is an equidistant code!) It follows that there is an exponent s such that $f(x) := x^s h(x)$ is the idempotent of H_m^*. Equation 14.2 now has the form

$$p_1(x)m_1(x) + x^s h(x) = 1.$$

Clearly $1 + f(x)$ is the idempotent of H_m. It follows that for any polynomial $q(x)$, the word $q(x)(1 + f(x))$ is in H_m.

In this chapter we shall consider cyclic codes for which the word length n is an *odd prime.* The alphabet will be $\mathbb{F}_q$, where q is required to be a *quadratic residue* (mod n). So we require that $q^{(n-1)/2} \equiv 1 \pmod{n}$. As usual α will denote a primitive n-th root of unity in an extension field of $\mathbb{F}_q$. Later we shall require that α satisfies one extra condition (only for convenience). Since the word length n is a prime, we can also consider the finite field $\mathbb{F}_n$ (this is *not* the alphabet). We partition this field into $\{0\}$, the *squares*, and the *nonsquares*:

$$R_0 := \{i^2 \pmod{n} : i \in \mathbb{F}_n, i \neq 0\},$$
$$R_1 := \mathbb{F}_n^* \backslash R_0.$$

The elements of R_0 are also called the *quadratic residues* mod n. Furthermore we define

$$g_0(x) := \prod_{r \in R_0} (x - \alpha^r), \quad g_1(x) := \prod_{r \in R_1} (x - \alpha^r).$$

Since we have required that $q \pmod{n}$ is in R_0, the polynomials $g_0(x)$ and $g_1(x)$ both have all their coefficients in $\mathbb{F}_q$. Furthermore

$$x^n - 1 = (x - 1)g_0(x)g_1(x).$$

(14.4) DEFINITION. The cyclic codes of length n over $\mathbb{F}_q$ with generators $g_0(x)$, respectively $(x - 1)g_0(x)$ are both called *quadratic residue codes* (QR codes).

For the time being we consider *extended quadratic residue codes* only for the case $q = 2$. The definition is as usual. In the binary case the code with generator $(x-1)g_0(x)$ consists of the words of even weight in the code with generator $g_0(x)$. If G is a generator matrix for the first of these codes, then $\begin{pmatrix} 1\,1 \cdots 1\,1 \\ G \end{pmatrix}$ is a generator for the second code and the generator for the extended code is $\begin{pmatrix} 1 & 1 \cdots 1\,1 \\ \mathbf{0}^\top & G \end{pmatrix}$ (cf. Chapter 9). Note that in the binary case the condition that q is a quadratic residue mod n is satisfied if $n \equiv \pm 1 \pmod 8$. If $j \in R_1$, then the permutation π_j maps R_0 into R_1 and vice versa. It is easily seen that if we replace R_0 by R_1 in (14.4) we obtain equivalent codes in the sense of Chapter 9. If $n \equiv -1 \pmod 4$ then $-1 \in R_1$ and therefore the transformation $x \mapsto x^{-1}$ maps a codeword of the code with generator $g_0(x)$ to a codeword of the code with generator $g_1(x)$.

(14.5) Theorem. *If $\mathbf{c} = c(x)$ is a codeword in the QR code with generator $g_0(x)$ and if $c(1) \neq 0$ and $w(\mathbf{c}) = d$, then*
(a) $d^2 \geq n$,
(b) if $n \equiv -1 \pmod 4$, then $d^2 - d + 1 \geq n$,
(c) if $n \equiv -1 \pmod 8$ and $q = 2$, then $d \equiv 3 \pmod 4$.

PROOF. Since $c(1) \neq 0$, the polynomial $c(x)$ is not divisible by $(x-1)$. By a suitable permutation π_j we can transform $c(x)$ into a polynomial $\hat{c}(x)$ that is divisible by $g_1(x)$ and of course again not divisible by $(x-1)$. This implies that $c(x)\hat{c}(x)$ is a multiple of $1 + x + x^2 + \ldots + x^{n-1}$. Since the polynomial $c(x)\hat{c}(x)$ has at most d^2 nonzero coefficients, we have proved the first assertion.

If $n \equiv -1 \pmod 4$, then in the proof of (a) we may take $j = -1$. In that case it is clear that $c(x)\hat{c}(x)$ has at most $d^2 - d + 1$ nonzero coefficients, proving (b).

Let $c(x) = \sum_{i=1}^{d} x^{l_i}$, $\hat{c}(x) = \sum_{i=1}^{d} x^{-l_i}$. If $l_i - l_j = l_k - l_l$, then $l_j - l_i = l_l - l_k$ and $l_i - l_k = l_j - l_l$. Hence, if terms in the product $c(x)\hat{c}(x)$ cancel, then they cancel four at a time. Therefore $n = d^2 - d + 1 - 4a$ for some $a \geq 0$, and (c) is proved. □

For the next two theorems we consider only binary QR codes. We define

$$\theta(x) := \sum_{r \in R_0} x^r. \tag{14.6}$$

The polynomial $\theta(x)$ is an idempotent because 2 is a quadratic residue mod n. Hence $\theta(\alpha)^2 = \theta(\alpha)$, that is, $\theta(\alpha) \in \mathbb{F}_2$. In the same way we see that $\theta(\alpha^i) = \theta(\alpha)$ if $i \in R_0$ and $\theta(\alpha^i) + \theta(\alpha) = 1$ if $i \in R_1$. We now make the extra restriction on α announced earlier, namely we require that $\theta(\alpha) = 0$. Then $\theta(\alpha^i) = 0$ if $i \in R_0$, $\theta(\alpha^i) = 1$ if $i \in R_1$. Finally, we observe that $\theta(\alpha^0) = (n-1)/2$. The following proposition is now obvious.

(14.7) Proposition. *If the primitive n-th root of unity α is suitably chosen, then the polynomial $\theta(x)$ of (14.6) is the idempotent of the QR code with generator $(x-1)g_0(x)$ if $n \equiv 1 \pmod 8$ and of the QR code with generator $g_0(x)$ if $n \equiv -1 \pmod 8$.* □

Let C be the circulant with the codeword θ as its first row. We define:

$$\mathbf{c} := \begin{cases} 0 & \text{if } n \equiv 1 \pmod 8, \\ 1 & \text{if } n \equiv -1 \pmod 8, \end{cases}$$

and

$$G := \begin{pmatrix} 1 & \mathbf{1} \\ \mathbf{c}^{\top} & C \end{pmatrix}.$$

It follows from Proposition (14.7) that the rows of G (which are clearly not independent) generate the extended binary QR code of length $n+1$.

We number the coordinate places for the extended binary QR code using the coordinates of the projective line, that is, $(1,0),(0,1),(1,1),\ldots,(n-1,1)$. Here the overall parity check is in front, in position (1,0). We now consider the permutations of the group PSL($2,n$) acting on the positions. This group is generated by the transformations $S := \left(\begin{smallmatrix} 1 & 1 \\ 0 & 1 \end{smallmatrix}\right)$ and $T := \left(\begin{smallmatrix} 0 & -1 \\ 1 & 0 \end{smallmatrix}\right)$. Since S leaves (1,0) invariant and acts as a cyclic shift on the remaining positions, it maps the code into itself. It is not difficult to show that T maps each row of G into a linear combination of at most three rows of G (see Exercise 5). These facts establish the following theorem.

(14.8) Theorem. *The automorphism group of the extended binary QR code of length $n+1$ contains* PSL($2,n$). □

REMARK. Gleason and Prange slightly altered the definition of an extended code for nonbinary QR codes. They require that the coordinate in the check position is multiplied by a (constant) factor in such a way that the resulting code is self-orthogonal if $n \equiv -1 \pmod 4$, respectively orthogonal to the other QR code if $n \equiv 1 \pmod 4$. They show that Theorem (14.8) then also holds in the nonbinary case. For a discussion see Assmus and Mattson (1969) or MacWilliams and Sloane (1977).

We now combine the results of Theorems (14.5) and (14.8).

(14.9) Theorem. *The minimum weight of the binary QR code of length n with generator $g_0(x)$ is an odd number d for which*
(a) $d^2 > n$ if $n \equiv 1 \pmod 8$,
(b) $d^2 - d + 1 \geq n$ if $n \equiv -1 \pmod 8$.

PROOF. The code is invariant under a doubly transitive permutation group. By the remark following Corollary 10.8 we see that the minimum distance must be odd. So, for a codeword $\mathbf{c}$ of minimum weight, the condition $c(1) \neq 0$ of Theorem (14.5) is satisfied. □

REMARK. The reader familiar with the theory of difference sets should have no difficulty checking that if equality holds in case (b) of Theorem (14.9), then there exists a projective plane of order $d-1$ (see the proof of (b) and (c) in Theorem (14.5)).

The bound of Theorem (14.9) is known as the *square root bound.* Assmus, Mattson, and Sachar generalized the theorem as follows. Let C be an $[n,(n+1)/2]$

cyclic code with minimum weight d. Assume that $C^\perp \subseteq C$ and that the words of minimum weight support a 2-design. Then $d^2 - d + 1 \geq n$ and equality holds iff the design is a projective plane of order $d - 1$.

We now give a first example, by far the most interesting QR code!

(14.10) Example. Over $\mathbb{F}_2$ we have

$$\begin{aligned} x^{23} - 1 &= (x-1)(x^{11} + x^9 + x^7 + x^6 + x^5 + 1)(x^{11} + x^{10} + x^6 + x^5 + x^4 + x^2 + 1) \\ &= (x-1)g_0(x)g_1(x). \end{aligned}$$

The binary QR code C of length 23 with generator $g_0(x)$ is a [23,12] cyclic code. By Theorem (14.9) the minimum distance d of this code is odd and satisfies $d^2 - d + 1 \geq 23$. So d is at least 7 and we already know that this implies that $d = 7$ and that C is the unique code $\mathcal{G}_{23}$.

The automorphism group of the binary Golay code is the Mathieu group M_{24}, a much larger group than the one guaranteed by Theorem (14.8) (but, as this proof shows, containing PSL(2, 23) as a subgroup).

The fact that the extended binary Golay code $\mathcal{G}_{24}$ is self-dual (see Chapter 11) also immediately follows from its representation as an extended QR code.

The following theorem is due to Assmus and Mattson (1969). It is one of the most important theorems in combinatorial coding theory.

(14.11) Theorem. *Let A be an $[n, k]$ code over $\mathbb{F}_q$ and let $B := A^\perp$ be the $[n, n-k]$ dual code. Let the minimum weights of these codes be d and e. Let t be an integer less than d. Let v_0 be the largest integer satisfying $v_0 - [(v_0 + q - 2)/(q-1)] < d$ and let w_0 be the largest integer satisfying $w_0 - [(w_0 + q - 2)/(q-1)] < e$, where if $q = 2$ we take $v_0 = w_0 = n$. Let B have at most $d - t$ non-zero weights less than or equal to $n - t$. Then for each weight v with $d \leq v \leq v_0$, the subsets of $S := \{1, 2, \ldots, n\}$ that support codewords of weight v in A form a t-design. Furthermore, for each weight w with $e \leq w \leq \min\{n - t, w_0\}$, the subsets of S that support words of weight w in B also form a t-design.*

Proof. We shall use the following notation. If T is a subset of S and C a code of length n, then $C \backslash T$ denotes the code that is obtained by *deleting* the coordinates in T from the codewords of C. The subcode of C consisting of the words of C that have a 0 in all the positions of T is denoted by $C_0(T)$. The proof is in six steps.

(1) By definition of v_0, two words of A with weight $\leq v_0$ and with the same support, must be scalar multiples of each other (since there exists a linear combination of these two words with weight $< d$). An analogous statement holds for B.

(2) Suppose that $(S, \mathcal{D})$ is a t-design and T a t-subset of S. Denote by α_i the number of blocks D of $\mathcal{D}$ with $|D \cap T| = i$. By Proposition (1.4) we have

$$\sum_{i=0}^{t} \binom{i}{j} \alpha_i = \binom{t}{j} \lambda_j \qquad (j = 0, 1, \ldots, t).$$

By solving these equations, we find that α_i does not depend on the choice of the set T. The statement for $i = 0$ implies that the complements of the blocks of $\mathcal{D}$ also form a t-design (see Proposition (1.39)).

(3) By definition of d, the deletion of any $d - 1$ columns of a generator matrix of A yields a matrix that still has rank k.

(4) Let T be a t-subset of S. By (3) the code $A\backslash T$ is an $[n - t, k]$ code. Clearly $C := B_0(T)\backslash T$ is a subcode of the dual of $A\backslash T$. Since the dimension of C is at least $n - k - t$, C must be $(A\backslash T)^{\perp}$. Let $0 < v_1 < v_2 < \ldots < v_r \leq n - t$ (where $r \leq d - t$) be the possible nonzero weights in the code B. Then these are also the only possible nonzero weights for the code C. Since the minimum weight of $A\backslash T$ is at least $d - t$, we know $d - t$ coefficients of the weight enumerator of $A\backslash T$. There are at most that many coefficients of the weight enumerator of C (the dual code) that we do not yet know. Theorem (9.14) yields a system of linearly independent equations for these unknowns. We could solve these equations. However, the important observation is that the solution, and hence the weight enumerator of the code C, does not depend on the choice of the set T. Using MacWilliams' Theorem a second time then shows that the same statement holds for the weight enumerator of $A\backslash T$.

(5) We now prove the second assertion of the theorem. Let $w \leq \min\{n - t, w_0\}$. Let $\mathcal{E}$ be the collection of w-subsets of S that support words of weight w in B. Consider the set $\mathcal{E}'$ of complements of sets in $\mathcal{E}$. For any t-subset T of S we find from (1) that the number of sets of $\mathcal{E}'$ containing T is $\frac{1}{q-1}$ times the number of words of weight w in C (as defined in (4)). By (4) this number does not depend on T. Hence $\mathcal{E}'$ is a t-design. By Proposition (1.39), the collection $\mathcal{E}$ is also a t-design.

(6) To prove the first assertion of the theorem, we start with $v = d$. Let $\mathcal{D}$ be the collection of d-subsets of S that support codewords of weight d in A. In the same way as in (5) we see that the number of sets in $\mathcal{D}$ containing a given t-subset T of S is $\frac{1}{q-1}$ times the number of words of weight $d - t$ in $A\backslash T$. By (4), this number does not depend on T. We now proceed by induction. Let $d \leq v \leq v_0$ and assume that the assertion of the theorem is true for all v' with $d \leq v' < v$. Let $\mathcal{D}$ be as before. The number of subsets of $\mathcal{D}$ containing a given t-subset T of S is $1/(q - 1)$ times the number of words of weight $v - t$ in $A\backslash T$ corresponding to words of weight v in A. By (4) the total number of words of weight $v - t$ in $A\backslash T$ does not depend on T. By the induction hypothesis and (2), the number of words of weight $v - t$ in $A\backslash T$ corresponding to words of weight $< v$ in A is also independent of T. Hence $\mathcal{D}$ is a t-design. $\square$

We shall now give several examples of this theorem.

(14.12) EXAMPLE. Take $n = 8$, $k = 4$, $q = 2$, and let $A = A^{\perp}$ be the extended [8,4] Hamming code. Then $d = e = 4$. Take $t = 3$. The condition of Theorem (14.11) is satisfied. Taking $v = 4$ we find the result of Proposition (9.12).

(14.13) EXAMPLE. Take $n = 12$, $k = 6$, $q = 3$, and let $A = A^{\perp}$ be the extended ternary Golay code. Then $d = e = 6$. (See Exercise 3.) Since all weights are divisible by 3, the condition of Theorem (14.11) is satisfied for $t = 5$. It follows that the supports of the words of weight 5 in the ternary Golay code form a 5-design.

(14.14) EXAMPLE. Let A be any $[12l, 6l]$ ternary self-dual code. Suppose $d > 3l$. Take $t = 5$. Since all weights in A are divisible by 3, the condition of the theorem is satisfied (as in the previous example). Taking $v = d$ we find that the supports of the minimum weight words of A form a 5-design.

(14.15) EXAMPLE. Take $n = 24$, $k = 12$, $q = 2$, and let $A = A^{\perp}$ be the extended binary Golay code $\mathcal{G}_{24}$. We have already proved that 0, 8, 12, 16, and 24 are the only weights that occur. Therefore the existence of the 5-(24,8,1) Steiner system also follows from Theorem (14.11).

(14.16) EXAMPLE. Let $n = 47$. Consider the binary QR code of length 47. By Theorem (14.9), the minimum distance d is at least 9. Then (c) of Proposition (14.5) shows that the minimum distance is at least 11. Since $|B(\mathbf{x}, 6)| > 2^{23}$, the code cannot be 6-error-correcting. Hence the minimum distance is 11. The extended code $\overline{C}$ is self-dual with minimum distance 12. MacWilliams' Theorem allows us to compute the weight enumerator of this code. The only weights that occur are 0, 12, 16, 20, 24, 28, 32, 36, and 48. (Note that this also follows from the fact that the code is doubly even and contains $\mathbf{1}$.) We can again apply Theorem (14.11) with $t = 5$. We may take $v =$12, 16, 20, or 24. In this way we find four different 5-designs (and also the complements).

These examples explain why so much effort has been put into finding the minimum distance of other quadratic residue codes. Very little is known (but it is doubtful whether long QR codes will yield interesting results). A survey for $n \leq 59$ can be found in Assmus and Mattson (1972a). Assmus and Mattson (1972b) proved that the minimum distance of the [60,30] extended binary QR code is 18 by using so-called *contractions*. These are mappings of self-orthogonal codes to shorter codes that are still self-orthogonal. Again Theorem (14.11) can be applied with $t = 5$ and one finds 5-designs on 60 points.

We give one more example of Theorem (14.11). It will turn up again in (16.20).

(14.17) EXAMPLE. Consider a $[2^{2l-1} - 1, 2^{2l-1} - 2l - 1, 5]$ primitive binary BCH code C (see (10.5)). As in previous examples one can use MacWilliams' Theorem to show that in $C^\perp$ no other weights occur than 2^{2l-2} and $2^{2l-2} \pm 2^{l-1}$. To apply Theorem 14.11 take $n = 2^{2l-1}$, $k = 2^{2l-1} - 2l - 1$, $q = 2$, and $A = \overline{C}$. In $A^\perp$ only three weights occur. So we can take $t = 3$ and find that for each v, the words of weight v support a 3-design.

From Proposition (12.7), we see that the dual code $C^\perp$ is a subcode of a punctured second order Reed–Muller code $\mathcal{R}(2, 2l-1)$. The codewords of this code are all the linear combinations of cyclic shifts of the idempotents of the codes with generator $(x^n - 1)/m_1(x)$, respectively $(x^n - 1)/m_3(x)$ (in the usual notation). Using this fact (see MacWilliams and Sloane (1977)) one can show that $C^\perp$ is a union of cosets of $\mathcal{R}(1, 2l - 1)$, where each coset corresponds to an alternating bilinear form of rank $2l - 2$. As in Lemma 12.9 this leads to the conclusion that all nonzero codewords in $C^\perp$ have weight $2^{2l-2} \pm 2^{l-1}$ or 2^{2l-2}.

REMARK. For $l = 2$, the extended code appeared in Chapter 12, Exercise 6.

In his thesis Delsarte (1973) generalized MacWilliams' relations and used the generalization to prove theorems similar to the Assmus–Mattson theorem. (The thesis contains much more than this! — Chapter 17 is based on it.) We shall present some of his theory and give some theorems relating (nonlinear) codes and designs.

In order to treat nonlinear codes we generalize the idea of a weight enumerator. We consider only *binary* codes.

(14.18) DEFINITION. Let C be a code of length n over $\mathbb{F}_2$. Define

$$A_i := |C|^{-1} \cdot |\{(\mathbf{x}, \mathbf{y}) : \mathbf{x} \in C, \mathbf{y} \in C, d(\mathbf{x}, \mathbf{y}) = i\}|.$$

We call $(A_0, A_1, \ldots, A_n)$ the *distance distribution* of C and define the *distance enumerator* by

$$A(x, y) := \sum_{i=0}^{n} A_i x^i y^{n-i}.$$

If C is linear, then A_i equals the number of words of weight i in C and $A(x, y)$ is the weight enumerator as defined in (9.13). Note that if C is linear and we replace C by $C - \mathbf{c}$, where $\mathbf{c} \in C$, then the weight enumerator does not change. Any code with this property is called *distance invariant*. For such a code the weight enumerator and the distance enumerator are the same.

We observe that $A(1, 1)$ equals the number of codewords. The following definition applies to polynomials, regardless of a connection with codes.

(14.19) DEFINITION. If $A(x, y) = \sum_{i=0}^{n} A_i x^i y^{n-i}$ is a polynomial in x and y (homogeneous of degree n), then

$$A'(x, y) := \frac{1}{A(1.1)} A(y - x, y + x)$$

is called the *MacWilliams transform* of $A(x,y)$. If C is linear and $A(x,y)$ is the weight enumerator of C, then $A'(x,y)$ is the weight enumerator of $C^\perp$ by Theorem (9.14). For any distance enumerator $A(x,y)$ we have $A_0 = 1$ and for any polynomial $A(x,y)$ with $A_0 = 1$ we have $A''(x,y) = A(x,y)$.

The following polynomials play an important rôle in coding theory.

(14.20) DEFINITION. Let n be fixed. The *Krawtchouk polynomial* $K_k(x)$ is defined by

$$K_k(x) := \sum_{j=0}^{k}(-1)^j \binom{x}{j}\binom{n-x}{k-j},$$

where

$$\binom{x}{j} := \frac{x(x-1)\cdots(x-j+1)}{j!}, \quad (x \in \mathbf{R}).$$

The following lemma is established by a trivial counting argument.

(14.21) Lemma. *If* $\mathbf{x} \in \mathbb{F}_2^n$ *has weight* i, *then*

$$\sum_{\substack{\mathbf{y}\in\mathbb{F}_2^n \\ w(\mathbf{y})=k}} (-1)^{\langle \mathbf{x},\mathbf{y}\rangle} = K_k(i).$$

The following inequalities are the basis of a bound known as the *linear programming bound* (cf. Van Lint (1982)). We shall use them for other purposes.

(14.22) Lemma. *Let* C *be a binary code of length* n *with distance distribution* $(A_0, A_1, \ldots, A_n)$. *Then*

$$\sum_{i=0}^{n} A_i K_k(i) \geq 0, \qquad k \in \{0, 1, \ldots, n\}.$$

PROOF. Let $|C| = M$. From Lemma 14.21 we find

$$\begin{aligned} M\sum_{i=0}^{n} A_i K_k(i) &= \sum_{i=0}^{n} \sum_{\substack{(\mathbf{x},\mathbf{y})\in C^2 \\ d(\mathbf{x},\mathbf{y})=i}} \sum_{\substack{\mathbf{z}\in\mathbb{F}_2^n \\ w(\mathbf{z})=k}} (-1)^{\langle \mathbf{z},\mathbf{x}-\mathbf{y}\rangle} \\ &= \sum_{\substack{\mathbf{z}\in\mathbb{F}_2^n \\ w(\mathbf{z})=k}} \left(\sum_{\mathbf{x}\in C}(-1)^{\langle \mathbf{z},\mathbf{x}\rangle}\right)^2 \\ &\geq 0. \end{aligned}$$

□

REMARK. If we replace the numbers A_i by the coefficients of the weight enumerator of C, then in the double sum over $\mathbf{x}$ and $\mathbf{y}$ we should take only the terms with $\mathbf{x} = \mathbf{0}$.

This amounts to leaving out the exponent 2 in the final expression. So, if in the proof we have equality, then this remains true after replacing the A_i by the coefficients of the weight enumerator. We use this fact below.

Let C be any binary code of length n with distance enumerator $A(x,y)$. If $A'(x,y) = \sum_{i=0}^{n} A'_i x^i y^{n-i}$ is the MacWilliams transform of $A(x,y)$, then

$$A'_k = \frac{1}{|C|}\sum_{i=0}^{n} A_i \sum_{j=0}^{k} (-1)^j \binom{i}{j}\binom{n-i}{k-j} \tag{14.23}$$
$$= \frac{1}{|C|}\sum_{i=0}^{k} A_i K_k(i).$$

So, by Lemma (14.22), the numbers A'_k are nonnegative. Clearly $A'_0 = 1$. We can now define what Delsarte called the four *fundamental parameters* of a code C. The first two are the *minimum distance* d and the number s of *distinct nonzero distances* between codewords (the number of nonzero A_i with $i > 0$). The other two are the *dual distance* d' defined by

$$d' := \min\{i \geq 1 : A'_i \neq 0\},$$

and the so-called *external distance* s' which is the number of nonzero coefficients A'_i with $1 \leq i \leq n$. The name 'dual distance' is clear because it is the distance of the dual code if C is linear. If C is linear, then the assertion of the following theorem is trivial; but for nonlinear codes it gives an interesting combinatorial interpretation to the dual distance.

(14.24) Theorem. *Let C be a binary code of length n with $|C| = M$ and dual distance d'. Let $[C]$ be the M by n array with the codewords of C as rows. Then if $r < d'$ any set of r columns of $[C]$ contains each r-tuple exactly $M/2^r$ times.*

PROOF. Since $A'_k = 0$ for $1 \leq k < d'$, we have from (14.22) (see the remark following the lemma) that $\sum_{\mathbf{x} \in C}(-1)^{\langle \mathbf{x}, \mathbf{z} \rangle} = 0$ for every $\mathbf{z} \in \mathbb{F}_2^n$ with $w(\mathbf{z}) = k$. Taking $w(\mathbf{z}) = 1$ we see that every column of $[C]$ must have $M/2$ ones and $M/2$ zeros. Then taking $w(\mathbf{z}) = 2$, we conclude that every pair of columns must contain each of the four possible pairs exactly $M/4$ times. Proceeding by induction the result follows. □

REMARK. The matrix $[C]$ mentioned in Theorem (14.24) is often referred to as an *orthogonal array of strength $d' - 1$*.

We shall now show a relation between these parameters and designs obtained from the codes. In the following $A_{i_1}, A_{i_2}, \ldots, A_{i_s}$ are the nonzero coefficients of $A(x,y)$.

An interesting situation arises when $s \leq d'$. In that case consider $A'(x,y)$ and substitute $y = 1$. We find

$$1 + \sum_{j=1}^{s} A_{i_j} y^{i_j} = \frac{|C|}{2^n} \sum_{i=0}^{n} A'_i (1-x)^i (1+x)^{n-i}.$$

Since $A'_i = 0$ for $1 < i < d'$, the l-th derivative of the right hand side in the point $x = 1$ does not depend on the values of the nonzero A'_i for $0 \le l \le s-1$. By calculating these derivatives we find s linearly independent equations for the numbers A_{i_j}. Therefore, these numbers depend only on the parameters n, $|C|$, and the numbers $i_1, i_2, \ldots, i_s$. Now, take any codeword as origin and consider the weight enumerator of C. Clearly, nonzero coefficients only occur for indices 0 and i_j $(1 \le j \le s)$. Using the remark following Theorem (14.22) we see that the MacWilliams transform of this weight enumerator must have 0 coefficients for $1 \le i \le d'$. Therefore the same equations that we found for the A_{i_j} apply. Then we have proved the following theorem.

(14.25) Theorem. *If C is a binary code for which $s \le d'$, then C is distance invariant.* □

If C is distance invariant, then $A_n = 0$ or $A_n = 1$. Let us assume that we know A_n. We shall show that one can obtain t-designs by considering the words of some fixed weight as blocks. We take $t := d' - \bar{s}$, where $\bar{s} := s - A_n$. Consider some fixed t-tuple of positions and denote by λ_{i_j} the number of code words of weight i_j that have ones in those t positions. Next, take $0 \le j \le \bar{s}$, $r := t + j$. We now count pairs $(\mathbf{c}, \mathbf{x})$, where $\mathbf{x}$ is a word of weight r with ones in the t fixed positions and $\mathbf{c}$ is a codeword with ones in all the positions of $\mathbf{x}$. From Theorem (14.22) we see that there are $\frac{M}{2^r}\binom{n-t}{j}$ such pairs (where $M := |C|$). On the other hand, the number of pairs is $\sum_{j=1}^{s}\binom{i_j-t}{j}\lambda_{i_j}$. In this way we find $\bar{s}$ linearly independent equations for the $\bar{s}$ unknowns λ_{i_j}. It follows that these numbers are independent of the choice of the t-tuple. So, we have the following theorem on designs from codes.

(14.26) Theorem. *Let C be a binary code for which $\bar{s} < d'$. If $i \ge d' - \bar{s}$ and $t := d' - \bar{s}$, then the words of weight i in C form a t-design.*

(14.27) Example. In Chapter 12 we showed that the Kerdock code $\mathcal{K}(m)$ had the fundamental parameters $s = 4$, $d' = 6$, and that $A_n = 1$ (where $n = 2^m$). So Theorem (14.26) can be applied with $t = 3$. We find Proposition (12.12).

Several authors have generalized the idea of QR codes. We mention Camion (1975), Delsarte (1971), and Ward (1974). An elementary presentation of the theory was given by Van Lint and MacWilliams (1978). We shall only sketch the ideas and show a connection with certain well known designs. In the generalization the length of the codes is $q = p^m$ $(m > 1)$. As alphabet we may use any finite field F. Let G be the additive group of F_q. We identify the positions of the code with elements of G and represent a codeword as

$$\mathbf{c} = \sum_{g \in G} c_g x^g, \qquad (c_g \in \mathsf{F}), \tag{14.28}$$

which is to be considered as a formal expression. Note that this corresponds to the representation of cyclic codes in the case that $m = 1$, where $G = \{0, 1, \ldots, p-1\}$.

Definition. The *group algebra* $\mathsf{F}G$ consists of all expressions of the form (14.28) with the following rules for addition and multiplication:

$$\sum a_g x^g \oplus \sum b_g x^g := \sum (a_g + b_g) x^g,$$

and

$$\sum a_g x^g * \sum b_g x^g := \sum \Big(\sum_{g_1+g_2=g} a_{g_1} b_{g_2} \Big) x^g,$$

where the summations are over all elements g in G. Then $(\mathsf{F}G, \oplus, *)$ is a ring.

Let ξ be a primitive p-th root of unity in some extension field $\hat{\mathsf{F}}$ of F. Let α be a primitive element of F_q. Every element $g \in G$ can be represented as $g = i_0 + i_1\alpha + \ldots + i_{m-1}\alpha^{m-1}$ with coefficients i_j in F_p. We now define the *character* $\psi_1 : G \to \hat{\mathsf{F}}$ by

$$\psi_1(g) := \xi^{i_0} \tag{14.29}$$

and for each $h \in G$ we define the character ψ_h by

$$\psi_h(g) := \psi_1(gh). \tag{14.30}$$

These characters are extended linearly to $\mathsf{F}G$ by defining

$$\psi_j \Big(\sum_{g \in G} a_g x^g \Big) := \sum a_g \psi_j(g).$$

The reader who is familiar with character theory will see that we now have all the characters of G and that $h \leftrightarrow \psi_h$ establishes an isomorphism between G and the group of characters. The proofs of all the facts that are stated below depend heavily on calculations with these characters. We leave out these (usually straightforward) proofs that can be found in Van Lint and MacWilliams (1978). In the following definition of the generalized codes U (respectively V) denotes the set of nonzero squares (respectively non-squares) in G.

(14.31) **Definition.** The *generalized quadratic residue code* of length q over F consists of all $\mathbf{c} = \sum c_g x^g$ such that $\psi_u(\mathbf{c}) = 0$ for all $u \in U$. We shall use the notation *GQR code.*

The code defined in (14.31) is denoted by A^+. The code B^+ is defined by replacing U by V. The codes A and B are the corresponding subcodes consisting of the words $\mathbf{c}$ satisfying the additional requirement $\psi_0(\mathbf{c}) = 0$. From the independence of the characters one sees that A^+ has dimension $\frac{1}{2}(q+1)$.

To see that we are indeed generalizing the earlier concept, the reader should convince himself that if $m = 1$ the definitions given in (14.31) and (14.4) are the same.

In the paper by Van Lint and MacWilliams it is shown that the idea of *idempotent* can be generalized. In the same way as was mentioned in the remark following (14.8) one can extend A^+ and B^+ to A_∞ and B_∞ by adding an extra symbol to the codewords such that the extended codes are dual codes. One can show that these codes are both invariant under PSL(2,q). It then follows that Proposition (14.5) also generalizes. In fact, one can show that if $m = 2l$ and $K := \mathbb{F}_{p^l}$, then the word $\mathbf{c} := \sum_{g\in K} x^g$ is in A^+. So, for even m the minimum weight is equal to $\sqrt{q}$.

In the paper quoted above a fairly difficult proof, restricted to the case $m = 2$, shows that the only words of minimum weight in A_∞ are the words in the orbit of $(\sum_{g\in K} x^g)$ under PSL(2,q), and multiples of these words. It then follows that the union of the supports of codewords of minimal weight in A_∞ and B_∞ are the images of $\mathbb{F}_p \cup \{\infty\}$ under the action of PGL(2,p^2) on the projective line of order p^2 (represented by $G \cup \{\infty\}$). So these words form the Möbius plane 3-$(p^2+1, p+1, 1)$, (see Chapter 1). Once again a design is obtained from the words of minimal weight in a code (in this case a code and its dual — for each of the codes A_∞ and B_∞, the words of minimum weight form a 2-design). Note that the partition of the blocks of the inversive plane of order 3, used in Chapter 6 (see Exercise 7, and the proof of (6.7)), is generalized here!

Van Lint and MacWilliams conjectured that this result on words of minimum weight was true for all even m. In order to prove this they needed the following theorem, which is quite interesting for its own sake. In 1984 Blokhuis gave an extremely elegant and ingenious proof of that theorem. We mention it since it is worth knowing and also because, combined with GQR codes, it yields new designs.

(14.32) Theorem. *Let q be a prime power and let S be a subset of the field $\mathbb{F}_{q^2}$ with the following properties:*
(a) The difference of any two elements of S is a square in $\mathbb{F}_{q^2}$;
(b) $0 \in S$ and $1 \in S$.
Then S is the subfield $\mathbb{F}_q$. □

Exercises

1. Let x^4+x+1 be the generator of the [15,11] binary Hamming code. Determine the idempotent of this code.

2. Let $x^n - 1 = g(x)h(x)$ over $\mathbb{F}_2$. Let C be the binary code with generator $g(x)$. Show that if $h(x)$ has even degree, the idempotent of C is $xg(x)h'(x)$.

3. Show that the ternary Golay code is a QR code. Show that the square root bound yields a better result than the BCH bound. Show that $d = 6$ and that the extended code is self-dual.

4. Determine all perfect QR codes.

5. Let G be the matrix introduced after Proposition (14.7) (that 'generates' the QR code). Let H be the matrix obtained by applying the permutation T to the coordinate positions. Consider the rows of G and H corresponding to position $(i, 1)$, where $i \in R_0$. Prove that the sum of these two rows is equal to the sum of the first two rows of G (corresponding to the positions (1,0) and (0,1)).

6. Show, using the methods of Chapter 9, that the extended ternary QR code of length 24 is self-dual. What is the square root bound for the minimum distance of the QR code? Show that Gleason and Prange's generalization of Theorem (14.8) implies that the extended code has minimum distance 9. Prove that the words of weight 9 in this code carry a 5-(24,9,6) design.

7. Show, using the same ideas as in the proof of the Assmus–Mattson theorem, that if C is a binary $[n, k, d]$ code ($k > 1$) such that for each weight $i > 0$ the codewords of weight i form a t-design ($t < d$), then the same is true for $C^{\perp}$.

8. Consider the field $\mathbb{F}_{3^2}$ with generator α satisfying $\alpha^2 = 1 + 2\alpha$. Go through the construction of the corresponding GQR code and then construct the 3-(10,4,1) design from the words of minimum weight in the code and its dual. (The solution is given in several examples in Van Lint and MacWilliams (1978)).

15. Symmetry codes over $\mathbb{F}_3$

In this chapter we treat a class of codes that have a number of properties in common with QR codes; for example, they also led to several new 5-designs. The results are due to V. Pless (1972, 1975). The construction of the ternary Golay code given in Proposition (11.9) is an example of these codes.

We remind the reader of the construction of Paley matrices. Let q be a power of an odd prime and let χ be the quadratic character on $\mathbb{F}_q$. We number the rows and columns of a matrix of order $q+1$ using the (inhomogeneous) coordinates of the projective line of order q, namely ∞ (for the point (1,0)) and the elements of $\mathbb{F}_q$. Define C_{q+1} by

$$c_{\infty,\infty} := 0, \quad c_{\infty,a} := 1, \quad c_{a,\infty} := \chi(-1)$$
$$c_{a,b} := \chi(b-a) \qquad (a, b \in \mathbb{F}_q).$$

So

$$C_{q+1} = \begin{pmatrix} 0 & \mathbf{1}^\top \\ \chi(-1)\mathbf{1} & S_q \end{pmatrix},$$

where S has entries $s_{a,b} = \chi(b-a)$.

We observe that
(a) $C_{q+1}C_{q+1}^\top = -I_{q+1}$ over $\mathbb{F}_3$ if $q \equiv -1 \pmod 3$;
(b) C_{q+1} is symmetric if $q \equiv 1 \pmod 4$;
(c) C_{q+1} is skew-symmetric if $q \equiv -1 \pmod 4$.

The Paley matrices are examples of a larger class of matrices known as conference matrices.

(15.1) DEFINITION. A *conference matrix* of order n is a real matrix C with entries ± 1 outside the diagonal and 0 on the diagonal such that $CC^\top = (n-1)I$.

The codes of this chapter are defined as follows.

(15.2) DEFINITION. Let $q \equiv -1 \pmod 6$. A *symmetry code* of dimension $q+1$ is a $[2q+2, q+1]$ ternary code Sym_{2q+2} with generator matrix

$$G_{2q+2} := (I_{q+1}\ C_{q+1}),$$

where C_{q+1} is a Paley matrix.

We observe that most of the following goes through if we only require that C_{q+1} is a (symmetric or skew-symmetric) conference matrix.

(15.3) Proposition. *The code* Sym_{2q+2} *is self-dual (and hence all its weights are divisible by 3).*

PROOF. This follows directly from property (a) of Paley matrices mentioned above. □

(15.4) Proposition. *The matrix*

$$G^*_{2q+2} := \left((-1)^{(q+1)/2}C_{q+1}\ I_{q+1}\right)$$

is also a generator matrix of Sym_{2q+2}.

PROOF. We have $G_{2q+2}G^*_{2q+2} = O$ because $C^{\mathsf{T}}_{q+1} = (-1)^{(q-1)/2}C_{q+1}$. Since Sym_{2q+2} is self-dual and the matrix G^*_{2q+2} has rank $q+1$, we are done. □

(15.5) Proposition. *The linear transformation of* $\mathbb{F}_3^{2q+2}$ *with matrix*

$$Z = \begin{pmatrix} O & I_{q+1} \\ (-1)^{q+1/2}I_{q+1} & O \end{pmatrix}$$

leaves Sym_{2q+2} *invariant.*

PROOF. This is a consequence of Proposition (15.4). □

(15.6) EXAMPLE. Take $q = 5$. The generator $G_{12} = (I_6\ C_6)$ is the matrix that was used in (11.9) to define the ternary extended Golay code $\mathcal{G}_{12}$. Several coding theorists independently found this representation of $\mathcal{G}_{12}$ but did not realize that the same construction for other values of q would yield interesting codes.

Before looking at other examples, we shall make some general assertions about the minimum distance of symmetry codes. In describing codewords we shall denote by $w_l(\mathbf{x})$, respectively $w_r(\mathbf{x})$, the contribution to the weight of $\mathbf{x}$ due to the first $q+1$, respectively the last $q+1$ coordinates.

(15.7) Lemma. *For every codeword* $\mathbf{x}$ *in a symmetry code of length* $2q+2$ *we have*
(a) if $w_l(\mathbf{x}) = 1$, *then* $w_r(\mathbf{x}) = q$,
(b) if $w_l(\mathbf{x}) = 2$, *then* $w_r(\mathbf{x}) = (q+3)/2$,
(c) if $w_l(\mathbf{x}) = 3$, *then* $w_r(\mathbf{x}) \geq 3(q-3)/4$.

PROOF. Consider three rows of the generator matrix. Since multiplication of a column by -1 does not alter weights, we may assume that $\mathbf{x}$ in (a), (b), (c) is the sum of one, two, or three rows of the following array.

$$
\begin{array}{ccccc ccc cccc}
 & & & & & & & & \overbrace{}^{a\text{ times}} & \overbrace{}^{b\text{ times}} & \overbrace{}^{c\text{ times}} & \overbrace{}^{d\text{ times}} \\
1 & 0 & 0 & \ldots & 0 & 0 & + & + & +\ldots+ & +\ldots+ & +\ldots+ & +\ldots+ \\
0 & 1 & 0 & \ldots & 0 & x_{21} & 0 & x_{23} & +\ldots+ & +\ldots+ & -\ldots- & -\ldots- \\
0 & 0 & 1 & \ldots & 0 & x_{31} & x_{32} & 0 & +\ldots+ & -\ldots- & +\ldots+ & -\ldots-
\end{array}
$$

Using the size and property (a) of the Paley matrix C_{q+1}, we find:

$$
\begin{aligned}
a+b+c+d &= q-2, \\
a+b-c-d &= -x_{23}, \\
a-b+c-d &= -x_{32}, \\
a-b-c+d &= -x_{21}x_{31}.
\end{aligned}
$$

Assertion (a) is obvious. To prove (b) we add the first two rows of the array and find

$$w_r(\mathbf{x}) = 2 + \tfrac{1}{2}(1 + x_{23}) + a + b = \tfrac{1}{2}(q+3).$$

Now add all three rows. Then

$$
\begin{aligned}
w_r(\mathbf{x}) &\geq b+c+d \\
&= \tfrac{1}{4}\{3(q-2) + x_{23} + x_{32} + x_{21}x_{31}\} \\
&\geq \tfrac{3}{4}(q-3).
\end{aligned}
$$

□

(15.8) Lemma. *Let w_1 and w_2 be integers. For a symmetry code we have*
(a) there is a codeword $\mathbf{x}$ with $w_l(\mathbf{x}) = w_1$ and $w_r(\mathbf{x}) = w_2$ iff there is a codeword $\mathbf{y}$ with $w_l(\mathbf{y}) = w_2$ and $w_r(\mathbf{y}) = w_1$;
(b) $w_r(\mathbf{x}) > 0$ for all codewords $\mathbf{x} \neq \mathbf{0}$.

PROOF. Property (a) is a consequence of Proposition (15.4). Property (b) follows from the fact that C_{q+1} is nonsingular. □

(15.9) EXAMPLE. We consider Sym_{36}, so $q = 17$. By Lemma 15.7 a codeword $\mathbf{x}$ with $w_l(\mathbf{x}) \leq 3$ has weight ≥ 12 (all weights are divisible by 3). We wish to show that there is no codeword of weight less than 12. By Lemma (15.8), it is sufficient to consider the possibility of a codeword $\mathbf{x}$ with $w_l(\mathbf{x}) = 4$, $w_r(\mathbf{x}) = 5$. All possibilities were checked by computer and no such combination was found. If one wishes to check this by hand, there are two ways to proceed. One way is to realize that C_{18} is a bordered circulant. This means that there are not too many essentially different ways to choose four rows. For example, if one of these is the top row, then without loss one chooses the next row and sees to it that the size of the largest gap between chosen rows occurs next. One could also use the method of the proof of Lemma (15.7) and add a fourth row to the array, asking whether the four could be orthogonal and have a sum with weight 4. There is in fact an essentially unique way of doing this: take $a = 3$, $b = c = d = 4$, and all $x_{ij} = 1$. Then the fourth row would be

$$0001\ldots0 +++ 0 + - + - - - + - - - - +++$$

and then one must show that this cannot be completed to a conference matrix. In this case the special form of the Paley matrix is not used. The case $q = 17$ is the largest for which hand calculation still is feasible. The result of this is that Sym_{36} has minimum distance 12.

We now apply the Assmus–Mattson Theorem (14.11). We have $n = 36$, $k = 18$, $q = 3$. Furthermore $d = e = 12$ and we must take $v_0 = w_0 = 23$. If $t = 5$, the conditions of the theorem are satisfied. We find 5-designs from the supports of codewords of weight 12, 15, 18, 21 and also the complementary designs. For all these designs (with one possible exception) there was no earlier construction.

In Pless (1975) the examples $q = 5$, $q = 11$, $q = 17$, $q = 23$, and $q = 29$ are treated. Of course $q = 5$ yields the well known Steiner system 5-(12,6,1) related to $\mathcal{G}_{12}$. All other 5-designs were new. The comparison with known designs was based on the study of automorphisms of the designs from symmetry codes.

Consider a matrix M with entries from $\mathbf{F}_q$. If replacing all nonzero entries by 1 yields a permutation matrix, then we call M a *monomial matrix*. We denote by $G(q)$ the group of all monomial matrices that map Sym_{2q+2} into itself (the mapping is $\mathbf{c} \mapsto \mathbf{c}M$). The corresponding group of permutation matrices is called $\overline{G}(q)$.

(15.10) Lemma. *If A and B are monomial matrices of order $q+1$ such that $A^{-1}C_{q+1}B = C_{q+1}$, then $\left(\begin{smallmatrix} A & O \\ O & B \end{smallmatrix}\right) \in G(q)$.*

PROOF. Codewords in Sym_{2q+2} have the form $\mathbf{a}(I_{q+1}\, C_{q+1})$, where $\mathbf{a}$ is a row vector of length $q+1$. Now

$$\mathbf{a}\,(I_{q+1}\, C_{q+1}) \begin{pmatrix} A & O \\ O & B \end{pmatrix} = \mathbf{a}\, A(I_{q+1} \quad A^{-1}C_{q+1}B),$$

which is clearly a codeword. □

Recall that the positions of C_{q+1} were numbered $\infty, 0, 1, \ldots, q-1$. We define analogues of the transformations S and T used in Chapter 14. As before, S is the permutation $x \mapsto x+1$ on the line (and S leaves ∞ invariant). We consider S and also C_{q+1} as linear transformations of $\mathbf{F}_3^{q+1}$. For the standard basis vectors $\mathbf{e}_\infty$, $\mathbf{e}_a$ ($a \in \mathbf{F}_q$) we have

$$\mathbf{e}_\infty C_{q+1} = \sum_{a \in \mathbf{F}_q} \mathbf{e}_a,$$

$$\mathbf{e}_a C_{q+1} = \chi(-1)\mathbf{e}_\infty + \sum_{b \in \mathbf{F}_q} \chi(b-a)\mathbf{e}_b.$$

It follows that

$$\mathbf{e}_\infty C_{q+1} S = \mathbf{e}_\infty S C_{q+1}$$

and

$$\begin{aligned}\mathbf{e}_a C_{q+1} S &= \chi(-1)\mathbf{e}_\infty + \sum_{b\in\mathbf{F}_q} \chi(b-a)\mathbf{e}_{b+1} \\ &= \chi(-1)\mathbf{e}_\infty + \sum_{b\in\mathbf{F}_q} \chi(b-a-1)\mathbf{e}_b \\ &= \mathbf{e}_a S C_{q+1},\end{aligned}$$

that is,

$$S^{-1} C_{q+1} S = C_{q+1}. \tag{15.11}$$

In the same way the permutation $P(b^2)$ defined by $x \mapsto b^2 x$ ($b \neq 0$) can be shown to satisfy

$$P(b^2)^{-1} C_{q+1} P(b^2) = C_{q+1}. \tag{15.12}$$

The permutation $Tx = x^{-1}$ (where $0^{-1} := \infty$) that we used for QR codes is now replaced by a monomial transformation that behaves in the same way as T when considered as a permutation only. We denote it by T^*. We define

$$\mathbf{e}_\infty T^* := (-1)^{(q-1)/2}\mathbf{e}_0, \quad \mathbf{e}_0 T^* := \mathbf{e}_\infty, \quad \mathbf{e}_a T^* := \chi(a)\mathbf{e}_{-1/a}\,(a \neq 0, \infty).$$

Once again it is straightforward to check that

$$(T^*)^{-1} C_{q+1} T^* = C_{q+1}. \tag{15.13}$$

The permutations S, $P(b^2)$, and the monomial transformation T^* generate a group R^*. It is easily seen that $R^*/\{I, -I\}$ is isomorphic to PSL(2,q). From Lemma (15.10) and the relations (15.11), (15.12), and (15.13) we then have the following theorem.

(15.14) Theorem. *The group $\overline{G}(q)$ contains a subgroup isomorphic to PSL(2,q).* □

REMARK. By Theorem (15.5), the monomial matrix

$$Z = \begin{pmatrix} O & (-1)^{(q+1)/2} I \\ I & O \end{pmatrix}$$

is also an automorphism of Sym_{2q+1}. The induced permutation $\overline{Z}$, which has order 2, commutes with those already found: so we have a subgroup PSL(2,q) $\times$ $\mathbf{Z}_2$ of $\overline{G}(q)$. In Pless (1975) it is shown that in fact $\overline{G}(q)$ contains a subgroup isomorphic to PGL$(2, q) \times \mathbf{Z}_2$. It is known that $\overline{G}(5)$ is M_{12}. In general, the group $G(q)$ is not known (see Exercise 5).

(15.15) EXAMPLE. From Exercise 6 of Chapter 14 we have a 5-(24,9,6) design. Assmus and Mattson (1969) have shown that the full automorphism group of this design is PSL(2,23). Using $q = 11$ we can construct the code Sym_{24} and from Lemma

15.7 we find the minimum distance to be 9. From this code we also find a 5-(24,9,6) design. By Theorem (15.14), this design has PSL(2,11) as a group of automorphisms. Since this is not a subgroup of PSL(2,23), the two designs are not the same. For more details we refer to Pless (1975); there the method of contraction mentioned in Chapter 14 is also used.

Exercises

1. Let C be a conference matrix of order n. Show that multiplying some columns and some rows by -1 produces another conference matrix. In this way we can normalize the matrix to make all entries in the top row (not on the diagonal) equal to 1. Show, using the same method as in the proof of Lemma (15.7), that one can make the normalized matrix symmetric if $n \equiv 2 \pmod 4$ and skew-symmetric if $n \equiv 0 \pmod 4$ (it is obvious that n must be even).

2. Let P be the Paley matrix C_{12}. Show that

$$H := \begin{pmatrix} 1 & 1 \\ -1 & 1 \end{pmatrix} \otimes (I + P)$$

is a Hadamard matrix. Use H to construct a skew-symmetric conference matrix C of order 24. With C construct a ternary self-dual [48,24] code. Find automorphisms of this code.

3. In Example (15.9) an assertion is made about a fourth row of the array of Lemma 15.7 that would yield a word of weight 9. Prove this assertion.

4. Let P be a symmetric matrix over $\mathbb{F}_3$ such that $P^2 = -I$. Let C be the code generated by $G := (I_n\, P)$ and let C^* be the code generated by $G^* := (P\, I_n)$. Prove that $\mathbb{F}_3^{2n} = C \oplus C^*$. (A special case is the fact that $\mathbb{F}_3^{12}$ is the direct sum of two codes that are equivalent to $\mathcal{G}_{12}$.)

5. Let $\overline{G}(q)$ be the automorphism group of the symmetry code Sym_{2q+2}, and H any subgroup.

(a) Prove that the converse of (15.10) holds: that is, if H preserves the first $q+1$ coordinate positions setwise, then its elements are all of the form described in (15.10). Deduce that only the identity permutation can fix the first $q+1$ coordinates elementwise.

(b) Prove that, if H preserves the partition of the set of coordinate positions into the orbits of $\overline{Z}$ (see Remark following (15.14)), then H commutes with $\overline{Z}$.

(c) Prove that, if H preserves neither the partition into the first and the last $q+1$ coordinates, nor the partition into the orbits of $\overline{Z}$, then H is doubly transitive.

(d) If you are familiar with permutation group theory, you might like to try to identify $\overline{G}(q)$ in all cases.

16. Nearly perfect binary codes and uniformly packed codes

We have seen (in Chapter 9, Exercise 8; and in Chapter 11) that words of fixed weight in a perfect code yield designs, in fact some of the most interesting ones! The two 5-designs connected with the Golay codes also have very interesting automorphism groups. These two facts explain the interest of both design-theorists and group-theorists in perfect codes. For both categories it is sad that Van Lint (1971) and Tietäväinen (1973) proved that if $e > 1$ and the size of the alphabet is a prime power, then there are no nontrivial perfect e-error-correcting codes except the Golay codes. Best (1982) and Hong (1984) extended this result to arbitrary alphabets for $e > 2$. So, only the cases $e = 1$ and non prime power alphabet with $e = 2$ remain open.

In this chapter we present some of the theory of binary *nearly perfect codes*, due to Goethals and Snover (1972). These codes are a special case of the class of *uniformly packed codes*, introduced by Semakov, Zinoviev and Zaitsev (1971). These codes also lead to t-designs. The theory of uniformly packed codes was further developed by Goethals and Van Tilborg. For a survey of most of what is known about these codes we refer to Van Tilborg (1976). Later in this chapter we shall describe some of the connections between uniformly packed codes, so-called *two-weight codes*, and strongly regular graphs. Much of the theory of uniformly packed codes depends heavily on algebraic methods; the non-existence results involve arguments from number theory. Since we are mainly interested here in combinatorial methods, we restrict our attention to results obtained by such methods.

First, we shall be concerned with the subclass of nearly perfect *binary* codes.

Let $C \subset \mathbb{F}_2^n$ be a code with minimum distance $d = 2t + 1$. For all $\mathbf{c} \in C$ we define

$$T(\mathbf{c}) := B(\mathbf{c}, t+1) \backslash B(\mathbf{c}, t).$$

We partition this set into two subsets $T_1(\mathbf{c})$ and $T_2(\mathbf{c})$, where

$$T_1(\mathbf{c}) := \{\mathbf{x} \in T(\mathbf{c}) : d(\mathbf{x}, C) = t\}.$$

So, the words in $T_2(\mathbf{c})$ have distance $t + 1$ to C.

(16.1) Lemma. *For each $\mathbf{c} \in C$ we have $|T_1(\mathbf{c})| \le \binom{n}{t}\lfloor\frac{n-t}{t+1}\rfloor$.*

PROOF. W.l.o.g., we can take $\mathbf{c} = \mathbf{0}$. A word $\mathbf{x}$ in $T_1(\mathbf{c})$ has weight $t+1$ and the (unique) codeword $\mathbf{u}$ at distance t to $\mathbf{x}$ has weight $2t+1$. For this word $\mathbf{u}$ there are exactly $\binom{2t+1}{t+1}$ words $\mathbf{x}$ in $T_1(\mathbf{c})$. Let N_{2t+1} denote the maximum number of codewords of weight $2t+1$ (given $\mathbf{0} \in C$). Count pairs $(\mathbf{u}, S)$, where $\mathbf{u}$ is a codeword of weight $2t+1$ and S is a t-subset of its ones. Doing this in two ways we find

$$N_{2t+1} \cdot \binom{2t+1}{t+1} \le \binom{n}{t}\lfloor\frac{n-t}{t+1}\rfloor.$$

This proves the lemma. □

(Note that the proof does not use the fact that $\mathbf{0} \in C$.)

The following theorem is a generalization of the (trivial) *sphere packing bound* given in Exercise 1 of Chapter 9. It is known as the *Johnson bound* (Johnson 1962).

(16.2) Theorem. *If $C \in \mathbb{F}_2^n$ is a code with minimum distance $d = 2t+1$, then*

$$|C| \cdot \left\{\sum_{i=0}^{n}\binom{n}{i} + \frac{1}{\lfloor n/(t+1)\rfloor}\binom{n}{t}\left(\frac{n-t}{t+1} - \lfloor\frac{n-t}{t+1}\rfloor\right)\right\} \le 2^n.$$

PROOF. Clearly the balls $B(\mathbf{c}, t)$, where $\mathbf{c}$ runs through C, are disjoint. The words in $R := \bigcup_{\mathbf{c}\in C} T_2(\mathbf{c})$ are not in any of these balls. A word $\mathbf{x} \in \mathbb{F}_2^n$ can be contained in at most $\lfloor n/(t+1)\rfloor$ distinct sets $T_2(\mathbf{c})$ (with $\mathbf{c} \in C$). From Lemma (16.1) we therefore find

$$|R| \ge |C| \cdot \left\{\binom{n}{t+1} - \binom{n}{t}\lfloor\frac{n-t}{t+1}\rfloor\right\} / \lfloor\frac{n}{t+1}\rfloor.$$

Since $|\mathbb{F}_2^n| = 2^n$ and $|B(\mathbf{c}, t)| = \binom{n}{t}$ the result follows. □

(16.3) DEFINITION. A binary code for which equality holds in the Johnson bound is called *nearly perfect.* Here we exclude codes for which $t+1$ divides $n+1$.

REMARK. Note that if $t+1$ divides $n+1$, then (16.2) reduces to the sphere packing bound and if equality holds, then we have a perfect code. This divisibility condition was one of the first known necessary conditions for the existence of a perfect code.

The following regularity condition is a consequence of the definition of nearly perfect codes.

(16.4) Proposition. *Let C be a nearly perfect binary code of length n and minimum distance $2t+1$. Let $\mathbf{x} \in \mathbb{F}_2^n$. Then*
(a) if $d(\mathbf{x}, C) > t$, there are exactly $\lfloor n/(t+1) \rfloor$ codewords $\mathbf{c}$ with $d(\mathbf{x}, \mathbf{c}) = t+1$.
(b) if $d(\mathbf{x}, C) = t$, then there are exactly $\lfloor (n-t)/(t+1) \rfloor$ codewords $\mathbf{c}$ with $d(\mathbf{x}, \mathbf{c}) = t+1$.

PROOF. Equality in the Johnson bound implies that equality occurs in (16.1) and in the estimate for N_{2t+1}. Both assertions follow directly from these equalities. □

The minimal number $\rho(C)$ such that every word in the space of all words has distance at most $\rho(C)$ to some codeword is called the *covering radius* of the code C. If a code C has minimum distance e and covering radius $e+1$, then C is called *quasi-perfect*. So, a nearly perfect code is quasi-perfect with the extra property that every word that has distance at least e to the code, has distance e or $e+1$ to the same number of codewords (namely $\lfloor n/(e+1) \rfloor$). We now consider a more general situation, again for quasi-perfect codes.

(16.5) Definition. An e-error-correcting code C of length n over $\mathbb{F}_q$ is called *uniformly packed* with parameters α and β iff for every word in $\mathbf{x} \in \mathbb{F}_q^n$
(a) if $\mathbf{x}$ has distance e to C, then $\mathbf{x}$ has distance $e+1$ to exactly α codewords, where $\alpha < (n-e)(q-1)/(e+1)$;
(b) if $\mathbf{x}$ has distance $> e$ to C, then $\mathbf{x}$ has distance $e+1$ to exactly β codewords.

For an explanation of the condition on α, see Exercise 2.

We shall give a few easy examples first. More complicated nearly perfect and uniformly packed codes will occur later.

(16.6) EXAMPLE. Consider the parity check matrix of a binary Hamming code (cf. Chapter 9) and delete any column. We obtain the parity check matrix of a linear code C with parameters $n = 2^m - 2$, $k = 2^m - m - 2$, and $d = 3$. If we substitute these parameters in (16.3), then equality holds, so C is a nearly perfect code.

(16.7) EXAMPLE. Let H be a Hadamard matrix of order 12. Take the 24 rows of H and $-H$ and then replace $+$ by 0 and $-$ by 1. We find a binary code of length 12 with 24 words. Clearly the minimum distance of this code is 6. We puncture this code to obtain a (11, 24, 5) code C. A word $\mathbf{z} \in \mathbb{F}_2^n$ can have distance 2 or 3 to at most four codewords. Suppose there is a word $\mathbf{z}$ that has distance 2 or 3 to four codewords. Without loss of generality we may assume that one of these is $\mathbf{0}$ and that $\mathbf{z}$ therefore has weight 2. The other words then have weight 5 and distance 6. In the original $\pm$ notation we have

$$\begin{aligned}
\mathbf{c}_1 &= ++ \quad +++ \quad +++ \quad +++,\\
\mathbf{c}_2 &= -- \quad --- \quad +++ \quad +++,\\
\mathbf{c}_3 &= -- \quad +++ \quad --- \quad +++,\\
\mathbf{c}_4 &= -- \quad +++ \quad +++ \quad ---.
\end{aligned}$$

This implies the existence of a Hadamard matrix of order 12 with the four rows $(+, \mathbf{c}_1)$, $(-, \mathbf{c}_2)$, $(-, \mathbf{c}_3)$, $(-, \mathbf{c}_4)$. Then $(-4, -4, -4, 0, 0, \dots, 0)$ is a linear combination of these four rows and must therefore be orthogonal to the remaining eight rows of the Hadamard matrix. This is clearly impossible. We have thus shown that a word $\mathbf{z}$ has distance 2 or 3 to at most three codewords.

Now count pairs $(\mathbf{c}, \mathbf{z})$, where $\mathbf{c} \in C$ and $\mathbf{z}$ has distance 2 or 3 to $\mathbf{c}$. Choosing $\mathbf{c}$ first, we find $24 \cdot (\binom{11}{2} + \binom{11}{3}) = 5280$ such pairs. The number of words $\mathbf{z}$ with distance 2 or 3 to the code is at most $2^{11} - 24 \cdot (1 + 11) = 1760$. Each has distance 2 or 3 to at most three codewords. It follows that every word $\mathbf{z}$ with distance > 1 to C has distance 2 or 3 to exactly three codewords. So C is uniformly packed with parameters $\alpha = 2$ and $\beta = 3$. This is an example of the situation that was excluded in the definition of nearly perfect codes, namely $e + 1$ divides $n + 1$.

We now look at designs obtained from nearly perfect codes.

(16.8) Theorem. *Let C be a nearly perfect binary code of length n with minimum distance $d = 2t + 1$. Let $\mathbf{0} \in C$. Then the words of weight d are the (characteristic functions of) blocks of a t-design $\mathcal{D}_1$ with parameters t-(n, d, λ), where $\lambda = \lfloor (n - t)/(t + 1) \rfloor$.*

PROOF. Let A be a t-subset of the coordinate places and let $\mathbf{a}$ be the word (of weight t) with 1's in the positions of A. By Proposition (16.4)(ii) there are $\lfloor (n - t)/(t + 1) \rfloor$ codewords $\mathbf{u}$ of weight d such that $d(\mathbf{u}, \mathbf{a}) = t + 1$, that is, such that A is contained in the block corresponding to $\mathbf{u}$. □

(16.9) Theorem. *Let C be a nearly perfect binary code of length n with minimum distance $d = 2t+1$. Let $\mathbf{0} \in C$. Then the collection $\mathcal{D}_2$ of $(t+1)$-subsets corresponding to words $\mathbf{u} \in T_2(\mathbf{0})$ (words of weight $t + 1$ with distance $t + 1$ to C) is a t-$(n, t + 1, \lambda)$ design, where $\lambda = (n - t) - (t + 1)\lfloor (n - t)/(t + 1) \rfloor$.*

PROOF. By definition, $\mathcal{D}_2$ consists of all the $(t + 1)$-subsets of the set of positions that are not contained in a block of $\mathcal{D}_1$. From Theorem (16.8) it follows that the $(t + 1)$-subsets that are contained in a block of $\mathcal{D}_1$ form a t-design. Apparently the blocks of $\mathcal{D}_2$ are all $(t + 1)$-sets which are not blocks of this design. (This is not the complement of the design!) □

(16.10) Theorem. *The design $\mathcal{D}_1$ of Theorem (16.8) can be extended to a $(t+1)$-$(n+1, t+2, \lambda)$ design, where $\lambda = \lfloor (n-t)/(t+1) \rfloor$.*

PROOF. Consider the extension $\overline{C}$ of the code C of Theorem (16.8). Fix k and delete the k-th coordinate of each word in $\overline{C}$. The resulting code C_k has length n, distance d, and $|C_k| = |C|$. It follows that C_k is nearly perfect (and contains $\mathbf{0}$). Since this is true for every k, it follows from Theorem (16.8) that the words $\mathbf{u}$ of weight $d+1$ in $\overline{C}$ are the blocks of a $(t+1)$-design with the same λ as in Theorem (16.8). □

(16.11) EXAMPLE. If we apply Theorem (16.10) to the shortened Hamming code of length 6, we find a 2-(7,4,2) design, which is of course the complement of the Fano plane.

The Preparata codes.

We shall now construct an infinite sequence of uniformly packed codes with minimum distance 5. The codes were first constructed by Preparata (1968). A very simple description of these codes was given by Baker, Van Lint, and Wilson (1983). We treat the extended codes first.

In the following m is odd ($m \geq 3$), $n = 2^m - 1$. Let F denote the field F_{2^m} and let $x \mapsto x^\sigma$ be an automorphism of F, that is, σ is a power of 2. We require that both $x \mapsto x^{\sigma+1}$ and $x \mapsto x^{\sigma-1}$ are one to one mappings. This is the case iff $(\sigma \pm 1, 2^m - 1) = 1$ and $\sigma = 2$ provides the easiest example.

For the admissible values of σ we shall define a code $\overline{\mathcal{P}}(\sigma)$ of length $2n+2 = 2^{m+1}$. The codewords will be described by pairs (X, Y), where $X \subset \mathsf{F}$, $Y \subset \mathsf{F}$. As usual we interpret the pair (X, Y) as the corresponding pair of characteristic functions, hencs as a (0,1)-vector of length 2^{m+1}. We shall let the zero element of F correspond to the first position (in the X-part and Y-part).

(16.12) DEFINITION. The extended *Preparata code* $\overline{\mathcal{P}}(\sigma)$ of length 2^{m+1} consists of the codewords described by all pairs (X, Y) satisfying

(a) $|X|$ is even, $|Y|$ is even,

(b) $\sum_{x \in X} x = \sum_{y \in Y} y$,

(c) $\sum_{x \in X} x^{\sigma+1} + \left(\sum_{x \in X} x\right)^{\sigma+1} = \sum_{y \in Y} y^{\sigma+1}$.

The code $\mathcal{P}(\sigma)$ is obtained by deleting the first coordinate.

For a discussion of the properties of these codes, we make the following conventions concerning notation. The symmetric difference of two sets X_1, X_2 is denoted by $X_1 \triangle X_2$ (this corresponds to addition of codewords). The set $\{x + \alpha : x \in X\}$ is denoted by $X + \alpha$. Many of the calculations depend on the following equality:

$$(a+b)^{\sigma+1} = a^{\sigma+1} + a^\sigma b + a b^\sigma + b^{\sigma+1}. \qquad (16.13)$$

The following theorems all hold for $\overline{\mathcal{P}}(\sigma)$. In some cases, we give the proof only for $\overline{\mathcal{P}} := \overline{\mathcal{P}}(2)$ because it is easier to read; the proof in the general case is similar.

(16.14) Proposition. *The code $\overline{\mathcal{P}}$ is distance invariant.*

PROOF. We compare a codeword (X_0, Y_0) with $(\emptyset, \emptyset) = \mathbf{0}$. Let $\alpha := \sum_{x \in X_0} x$. The mapping $(X, Y) \mapsto (U, V)$, where $U := (X \triangle X_0) + \alpha$, $V := (Y \triangle Y_0)$ is clearly one to one. We now show that if (X, Y) is a codeword, then so is (U, V) and vice versa. The conditions (a) and (b) of (16.12) are trivially checked. For condition (c), we use (16.13):

$$\begin{aligned}\sum_{x \in U} x^3 + \Big(\sum_{x \in U} x\Big)^3 &= \sum_{x \in X} (x+\alpha)^3 + \sum_{x \in X_0} (x+\alpha)^3 + \Big(\sum_{x \in X} x + \alpha\Big)^3 \\ &= \sum_{x \in X} x^3 + \sum_{x \in X_0} x^3 + \Big(\sum_{x \in X} x\Big)^3 + \alpha^3 \\ &= \sum_{y \in Y} y^3 + \sum_{y \in Y_0} y^3 = \sum_{y \in V} y^3.\end{aligned}$$

□

The proofs of the main properties of these codes become simpler if we first find some automorphisms of the codes.

(16.15) Proposition. *The group* Aut $\overline{\mathcal{P}}(\sigma)$ *contains the permutations*

(a) $(X, Y) \mapsto (X + c, Y + c)$, $\quad c \in \mathsf{F}$,

(b) $(X, Y) \mapsto (Y, X)$,

(c) $(X, Y) \mapsto (\alpha X, \alpha Y)$, $\quad \alpha \in \mathsf{F}^*$.

(d) $(X, Y) \mapsto (X^\phi, Y^\phi)$, $\quad \phi \in \mathrm{Aut}(\mathsf{F})$.

PROOF. To check condition (c) of (16.12) for the mapping (a), again use (16.13). All other properties are trivially true. □

(16.16) Proposition. $\overline{\mathcal{P}}$ *has minimum distance 6.*

PROOF. By Proposition (16.14) it is sufficient to show that the minimum weight is 6. Since there are obviously no words of weight 2, we only have to show that weight 4 cannot occur. First assume that there is a codeword $(\{x_1, x_2\}, \{y_1, y_2\})$. By Proposition (16.15) we may then assume that $x_1 = 0$. Then (c) of (16.12) yields

$$y_1^3 + y_2^3 = 0,$$

and this implies that $y_1 = y_2$, a contradiction.

By Propositions (16.14) and (16.15), it remains to check the possibility of $|X| = 4$, $Y = \emptyset$, where $X = \{0, a, b, c\}$. From (b) and (c) of (16.12) we find

$$a + b + c = 0,$$
$$a^3 + b^3 + c^3 = 0.$$

Substitution of the first of these in the second and then using formula 16.13 yields $ab(a+b)=0$, whence $a=b$, a contradiction.

Finally we show that there are indeed codewords of weight 6. Let a, b, c be distinct elements of $\mathbb{F}$. Define y by $y^3 := a^3+b^3+c^3$ and define x by $x := a+b+c+y$. Then $(\{0, x\}, \{a, b, c, y\})$ is a codeword (note that $x \neq 0$). □

From Definition (16.12), it is not trivial to see what the number of codewords in a Preparata code is. We will need results from Chapter 10.

(16.17) Proposition. $|\overline{\mathcal{P}}(\sigma)| = 2^k$ *where* $k = 2^{m+1} - 2m - 2$.

PROOF. Consider (16.12). A set X satisfying (a) can be chosen in 2^n ways. We now count how many sets Y in $\mathbb{F}^*$ satisfy (b) and (c), and adjoin the element 0 to each such set if necessary to satisfy condition (a). Let ω be a primitive element of $\mathbb{F}$ and let $m_i(x)$ be the minimal polynomial of ω^i. We have two equations over $\mathbb{F}$ for the elements y of Y (namely (16.12) (b) and (c)). If we consider $\mathbb{F}$ as an m-dimensional vector space over $\mathbb{F}_2$, then these equations become $2m$ *linear* equations. We claim that these linear equations are independent. This is so because $(\sigma+1, n) = 1$ and hence $m_{\sigma+1}(x)$ has degree m, which implies that the cyclic code over $\mathbb{F}_2$ with length n and generator $m_1(x)m_{\sigma+1}(x)$ has dimension $n-2m$. From this it follows that for each choice of X, the equations (16.12) (b) and (c) have 2^{n-2m} solutions Y with $Y \subset \mathbb{F}^*$. This proves the assertion. □

We have now achieved our goal.

(16.18) Theorem. *The code $\mathcal{P}(\sigma)$ is a nearly perfect binary code of length $2^{m+1}-1$ with minimum distance 5.*

PROOF. This follows from the definition of nearly perfect codes and Propositions (16.16) and (16.17). □

The Preparata codes have a number of remarkable properties (besides being nearly perfect). The fact that $\mathcal{P}(\sigma)$ is nearly perfect determines the weight enumerator of $\overline{\mathcal{P}}(\sigma)$. As was already remarked in Chapter 12, it turns out that this code and the Kerdock code of the same length are *formally dual*, in the sense that they satisfy the MacWilliams relations!

The Preparata code of length 16 is unique; it is in fact the Nordstrom–Robinson code (Chapter 11, Exercise 2). So it is equal to the Kerdock code $\mathcal{K}(4)$.

It is not difficult to show (using (16.4)) that the Preparata code is a subcode of the Hamming code of the same length. (See Exercise 4.) One can also show (see Exercise 7) that the extended Hamming code is a union of translates of a Preparata code. Again, these properties are in some sense 'dual' to those of Kerdock codes, viz., a Kerdock code is a union of translates of the first order Reed–Muller code (the dual of the extended Hamming code).

The codes $\overline{\mathcal{P}}(\sigma)$ provide an interesting application of Theorem (16.10).

(16.19) EXAMPLE. Let $m = 2k - 1$. Consider the extended Preparata code of length 4^k. By Theorem (16.10), the words of weight 6 in this code form a 3-$(4^k, 6, (4^k - 4)/3)$ design. We consider the special case $k = 2$. (We take $\sigma = 2$.) The reader should check (see Exercise 8) that the codewords for which $\sum_{x \in X} x = 0$ form a linear subcode with minimum distance 8. This subcode has seven cosets and each of these has 16 words of weight 6. Since these words mutually have distance at least 6, they must form the blocks of a 2-(16,6,2) design (corresponding to a Hadamard matrix). If we start with one of these designs, then the blocks of the 3-(16,6,4) design are obtained by simultaneous cyclic shifts on the positions (in the X-part and Y-part) corresponding to nonzero elements of $\mathbb{F}_{2^3}$.

Note that the blocks of the 3-design form a constant weight code with weight 6, distance 6, and with 112 words (see Exercise 6).

It was shown by Lindström (1975) and Van Tilborg (1976) that the shortened Hamming codes and the Preparata codes are the only binary nearly perfect codes; (more precisely, such codes must have the parameters of these codes). In fact, Van Tilborg proved that for $e \geq 4$ there does not exist a nontrivial uniformly packed code. For smaller values of e there are many interesting examples.

As a preparation for the next theorem, we give an example of a uniformly packed code for which this property follows from design properties of the codewords.

(16.20) EXAMPLE. In Example (14.17), we saw that the words of weight 6 in the extension $\overline{C}$ of the 2-error-correcting binary BCH code C of length $2^{2l-1} - 1$ form a 3-design. Consider a word $\mathbf{x}$ of weight 2 or 3. Then $\mathbf{x}$ has distance 2 or 3 to $\mathbf{0}$ and distance at least 2 to all other codewords. Using the 3-design, we see that $\mathbf{x}$ must have distance 2 or 3 to exactly λ codewords, where λ is the parameter of the 3-design. Therefore C is a uniformly packed code with parameters $\alpha = \lambda$, $\beta = \lambda + 1$. The reader should check for himself that this code is not nearly perfect.

The application of the Assmus–Mattson theorem in Example (14.17) was based on the fact that the dual code had only three weights (not 0). Goethals and Van Tilborg (see Van Tilborg (1976)) proved that it is this property that makes the code uniformly packed. The proof uses Delsarte's theory and exploits properties of Krawtchouk polynomials extending methods that we sketched in Chapter 14. It would take too long to give the proof here, so we only state the theorem.

(16.21) Theorem. *Let C be an e-error-correcting linear code. Then C is uniformly packed iff the number of nonzero weights in $C^{\perp}$ is $e+1$.* □

In fact, they show a more general result, namely that C is uniformly packed if its external distance is $e+1$.

A special case that has been extensively studied is $e=1$. In this case $C^{\perp}$ has only two nonzero weights. Such codes are called *two-weight codes.* We shall give a few examples of such codes. This list is far from complete; see Calderbank and Kantor (1986) for many further examples.

First, recall Exercise 12 of Chapter 9: if a linear code has minimum distance 3, then the columns of a parity check matrix represent different points in projective space. This has led to the name *projective code* for a code for which the generator matrix has pairwise linearly independent columns (that is, for which the dual code has distance at least 3). We give below several examples of sets of points in projective spaces which meet every hyperplane in a or b points, for some numbers a and b.

There is a strong connection between two-weight codes and one of our earlier topics, namely strongly regular graphs. This was shown by Delsarte (1972) with the next theorem and its converse ((16.22) and (16.31)).

(16.22) Theorem. *Let C be a q-ary two-weight projective code of length n. Let the two weights be w_1, w_2 ($w_1 < w_2$). Define the graph $\Gamma(C)$ by taking the codewords as vertices and joining $\mathbf{x}$ and $\mathbf{y}$ if $d(\mathbf{x},\mathbf{y}) = w_1$. Then $\Gamma(C)$ is a strongly regular graph.*

PROOF. Let k be the dimension of C and let there be b_i words of weight w_i ($i=1,2$). Consider the matrix $A = \binom{A_1}{A_2}$, in which the rows of A_i are all the codewords of weight w_i ($i=1,2$). Then every column of A has $q^k - q^{k-1}$ nonzero entries. Therefore we have

$$b_1 + b_2 = q^k - 1$$
$$b_1 w_1 + b_2 w_2 = n(q^k - q^{k-1}).$$

So we can calculate b_1 and b_2.

Now take any column of A and let r_i denote the number of zeros in this column in A_i ($i=1,2$). The words with a zero in this column form a two-weight (projective) subcode of C and we can therefore calculate r_1 and r_2 in the same way that we calculated b_1 and b_2. This implies that these numbers do not depend on the column that was chosen.

We now know that every column of A_1 has r_1 zeros. Therefore we know the sum of the distances of the rows of A_1 to the first row of A_1. Let s_i of these words have distance w_i to the first row of A_1 ($i=1,2$). Then the sum of the distances to the first row is $s_1w_1 + s_2w_2$ and of course $s_1 + s_2 = b_1 - 1$. Therefore we can also calculate s_1 and s_2.

We have thus proved that if $\{\mathbf{x}, \mathbf{y}\}$ is an edge in $\Gamma(C)$, then there is a constant number of vertices $\mathbf{z}$ joined to $\mathbf{x}$ and to $\mathbf{y}$. The calculation in the case that $\{\mathbf{x}, \mathbf{y}\}$ is not an edge is similar. □

There is another construction of strongly regular graphs from uniformly packed codes, as follows. Let C be a linear uniformly packed 1-error-correcting q-ary code. We take the vertices of the graph $\Gamma = \Gamma^*(C)$ to be the cosets of C in $V = \mathbb{F}_q^n$. Two vertices are adjacent if the cosets contain words with Hamming distance 1 from each other. Clearly the graph admits the translation group of V/C as an automorphism group.

(16.23) Theorem. *Let C be a q-ary $[n, d, \geq 3]$ uniformly packed code, with parameters α and β. Then the graph constructed above is strongly regular, with parameters $(q^{n-d}, n(q-1), q-2+2\alpha, 2\beta)$.*

PROOF. Using the translation group, we may assume that an arbitrarily chosen vertex x contains the zero vector (hence is the coset C). The neighbours of x are the cosets containing vectors of weight 1; since no two such vectors lie in the same coset, the valency is $n(q-1)$.

Suppose that y is adjacent to x, and y contains a vector $\mathbf{y}$ with support $\{i\}$. The $q-2$ multiples of $\mathbf{y}$ other than $\mathbf{0}$ and $\mathbf{y}$ lie in cosets adjacent to x and y. Any further common neighbour contains a vector with support $\{j\}$, say, where $\{i, j\}$ is contained in the support of a word $\mathbf{w}$ of C of weight 3 with $d(\mathbf{y}, \mathbf{w}) = 2$. There are α such words $\mathbf{w}$, and for each α there are two choices for $\mathbf{y}$. So $\lambda = q - 2 + 2\alpha$.

The argument for non-adjacent vertices is similar. □

We now proceed to give examples of sets in projective spaces with just two cardinalities of hyperplane sections. The reader should calculate, for at least some of these, the parameters of the graphs and codes which arise.

(16.24) EXAMPLE. In a projective line, hyperplanes are points, and so every set has this property, with $a = 0$, $b = 1$. The corresponding graphs are of Latin square type.

(16.25) EXAMPLE. The complement of a k-dimensional subspace S of $\mathrm{PG}(n, q)$ meets a hyperplane containing S in $q^{k+1}(q^{n-k} - 1)/(q-1)$ points, and meets any other hyperplane in $q^{k+2}(q^{n-k-1} - 1)/(q-1)$ points, since such a hyperplane meets S in a $(k-1)$-dimensional subspace.

(16.26) EXAMPLE. We saw some examples in $\mathrm{PG}(2, q)$ in Chapter 7. In (7.14), Denniston's construction of sets meeting every line in 0 or k points was mentioned, where q is a power of 2, and k any divisor of q. In (7.18), we met Baer subplanes and unitals, two classes of subsets of $\mathrm{PG}(2, q^2)$ meeting every line in 1 or $q+1$ points (and having $q^2 + q + 1$ and $q^3 + 1$ points respectively).

(16.27) EXAMPLE. Another example is an *ovoid* in PG(3,q), a set of q^2+1 points, no three on a line, and such that each hyperplane meets the ovoid in 1 point or in $q+1$ points. We met ovoids in connection with inversive planes in Chapter 1. Two classes of examples are known: the elliptic quadrics, and the Suzuki-Tits ovoids (the latter occurring only for $q = 2^{2h+1}$, $h > 1$).

(16.28) EXAMPLE. Let X be a non-singular quadric in PG$(n-1,q)$, with n even. A section of X by a non-tangent hyperplane is a non-singular quadric in PG$(n-2,q)$; all such quadrics are equivalent, since $n-2$ is odd (see Dickson (1958)), and so have the same cardinality. Any section by a tangent hyperplane has the form $X \cap x^\perp$ for some $x \in X$; all such sets have the same cardinality. So X has the required property. Note that there are two types of non-singular quadric in PG$(n-1,q)$ with n even, the elliptic and the hyperbolic. Elliptic quadrics in PG$(3,q)$ are ovoids, so this example overlaps with the preceding one. We now work a particular case in detail.

Let W be the set of 35 points $\mathbf{x}$ in $\mathbb{F}_2^6\backslash\{\mathbf{0}\}$ on the quadric with equation $x_1x_2 + x_3x_4 + x_5x_6 = 0$. We take these vectors as columns of a 6 by 35 matrix G. We claim that G generates a two-weight code. To see this, we first observe that the i-th row of G is the characteristic function of the intersection of W and the hyperplane with equation $x_i = 1$ $(1 \le i \le 6)$. Hence the weight of a linear combination $\mathbf{a}^\top G$ $(\mathbf{a} \in \mathbb{F}_2^6)$ is the number of solutions of

$$x_1x_2 + x_3x_4 + x_5x_6 = 0 \quad \text{and} \quad \sum_{i=1}^{6} a_ix_i = 1.$$

Without loss of generality we may take $a_1 = 1$ (unless $\mathbf{a} = \mathbf{0}$). By substitution and the affine transformation

$$y_2 = x_2, \qquad y_3 = x_3 + a_4x_2, \qquad y_4 = x_4 + a_3x_2,$$

$$y_5 = x_5 + a_6x_2, \qquad y_6 = x_6 + a_5x_2,$$

(which is invertible) we see that we must count the number of solutions of the equation

$$(1 + a_2 + a_3a_4 + a_5a_6)y_2 + y_3y_4 + y_5y_6 = 0.$$

If the coefficient of y_2 is 1, then the number of solutions is 16 and otherwise it is 20. So we indeed have a two-weight code of length 35 and it is projective. The dual of the code generated by G is therefore a uniformly packed code. From this code we find a strongly regular graph on 64 points. (Compare Example (5.16).)

The following example is slightly more difficult. It is related to many other areas of combinatorics and sometimes called the Hill cap (cf. Hill (1978)).

(16.29) EXAMPLE. Let $\mathcal{D}$ be a 2-(6,3,2) design. We define a set S^* of vectors in $\mathbb{F}_3^6$ as follows. A vector $\mathbf{x}$ is in S^* if all its coordinates are nonzero and an even number of them is -1; furthermore $\mathbf{x}$ is in S^* if it supports a block of $\mathcal{D}$. Let S be the set of points of PG(5,3) corresponding to the points of S^*. Then $|S| = 4 \cdot 10 + 2^4 = 56$. Using the fact that the complement of a block of $\mathcal{D}$ is not a block, it is not difficult to show that each hyperplane of PG(3,5) meets S in 11 or in 20 points. Using this set (the Hill cap) as in the construction described above, now deleting the columns not corresponding to points of S, we find a ternary code of length 56 with two weights, namely 36 and 45. So, the dual code is uniformly packed.

(16.30) EXAMPLE. Let C be the [5,4] binary even-weight code. This is a two-weight code with weights 2 and 4. Hence $\Gamma(C)$ is a strongly regular graph on 16 points with parameters (16,5,0,2). It is the Clebsch graph. This definition is in fact the same as was given in Chapter 2.

The next theorem is the converse to Theorem (16.22). It is more difficult to prove. For the proof we refer to Delsarte (1972).

(16.31) Theorem. *Let Γ be a strongly regular graph on $n = p^a$ vertices (p a prime). Let k, r, s be the eigenvalues of Γ (all integers). Let the elementary abelian p-group G_n be a regular group of automorphisms of Γ. Then there is a two-weight code C over $\mathbb{F}_p$ such that $\Gamma = \Gamma(C)$.* □

The proof of the theorem also shows that the word length N of C, and the weights w_1, w_2 of C are related by

$$(r-s)(p-1)N = -k - s(p^a - 1),$$

and

$$(r-s)w_i = \left(\frac{-s+(-1)^i-1}{2}\right)p^{a-1}, \quad (i = 1,2).$$

(16.32) EXAMPLE. Consider the partial geometry with parameters (5,5,2) of Van Lint and Schrijver, treated in Chapter 7. Corresponding to this partial geometry there is a strongly regular graph. The easiest way to describe this graph is to take as its vertices the codewords of the [5,4] ternary linear code C defined by $\mathbf{c} \in C \Leftrightarrow \langle \mathbf{c}, \mathbf{1} \rangle = 0$ and to join two vertices by an edge iff their distance is 2 or 5. It is easily checked that this is indeed a strongly regular graph with parameters (81,30,9,12) and that the lines of the partial geometry are certain cliques of the graph. For our present application we need only to observe that the graph clearly has an elementary abelian group of order 81 (namely the translations) as a group of automorphisms. By Theorem (16.31) this graph is associated with a two-weight code of dimension 4 over $\mathbb{F}_3$. The formulas given after the theorem show that this code has length 25 and that the two weights are 15 and 18. By Theorem (16.21) the dual of this code is a uniformly packed code.

Exercises.

1. Denote by $A(n, d, w)$ the maximum number of codewords in a binary code of length n and minimum distance $\geq d$ for which all codewords have weight w. Prove

that

$$A(n, 2k-1, w) = A(n, 2k, w) \le \lfloor \frac{n}{w} \lfloor \frac{n-1}{w-1} \lfloor \dots \lfloor \frac{n-w+k}{k} \rfloor \dots \rfloor \rfloor \rfloor.$$

Use this result to give a proof of Lemma 16.1.

2. Let C be a q-ary e-error-correcting uniformly packed code of length n with parameters α and β. Show that

$$|C| \left\{ \sum_{i=0}^{e-1} \binom{n}{i} (q-1)^i + (1 - \frac{\alpha}{\beta}) \binom{n}{e} (q-1)^e + \frac{1}{\beta} \binom{n}{e+1} (q-1)^{e+1} \right\} = q^n.$$

Check that $\alpha = (n-e)(q-1)/(e+1)$ would imply that C is perfect.

3. Consider the following construction (analogous to (16.12)). Binary words of length 24 are described by triples (X, Y, Z), where X, Y, Z are subsets of $\mathbb{F}_8$. We require

(a) $\sum_{x \in X} x = \sum_{y \in Y} y = \sum_{z \in Z} z$;
(b) $\sum_{x \in X} x^3 + \sum_{y \in Y} y^3 + \sum_{z \in Z} z^3 = 0$.

Show that the code defined in this way is the binary Golay code. (This construction is due to G. Glauberman.)

4. Consider a Preparata code $\mathcal{P}(2)$ and adjoin to this code all the words that have distance 3 to the code. Show that the new code is the Hamming code of the corresponding length.

5. Determine the number of words of weight 5, respectively 6, in the code $\mathcal{P}(2)$ of length 15.

6. Let C be a constant weight code with weight 6 and distance 6. If $|C| = 112$, then show that the words of C are the blocks of a 3-design.

7. Let C_0 be the extended Preparata code $\overline{\mathcal{P}}(\sigma)$. For each $\alpha \in \mathbb{F}^*$ we define the code C_α to be the code obtained by adding the word corresponding to $(\{0, \alpha\}, \{0, \alpha\})$ to the codewords of C_0. Prove the following.

(a) For each $\alpha \in \mathbb{F}^*$, the code C_α has minimum weight 4.
(b) The codes C_α $(\alpha \in \mathbb{F})$ are pairwise disjoint.
(c) The code $\overline{H} := \bigcup_{\alpha \in \mathbb{F}} C_\alpha$ is linear.
(d) $\overline{H}$ is the extended Hamming code.

8. Consider the Preparata code $\mathcal{P}(2)$ of length $2n + 1 = 2^{m+1} - 1$ as defined in (16.12). We define the linear code C_m to be the subcode of $\mathcal{P}(2)$ consisting of the words with $\sum_{x \in X} x = 0$. Using the notation of cyclic codes we now denote codewords of C_m as $(c_1(\xi), i, c_2(\xi))$ (as usual. the polynomials are in the ring $\mathbb{F}_2[\xi]/(\xi^n - 1)$.) In our case $c_1(\xi)$ is the representation of the word corresponding to $X \backslash \{0\}$, $i = 1$ if $0 \in Y$ and 0 otherwise, and $c_2(\xi)$ represents $Y \backslash \{0\}$.

(a) Show that the definition of C_m implies that $c_1(\omega) = 0$, where ω is a primitive element of $\mathbb{F}_{2^m}$.

(b) Show that conditions (a) to (c) of (16.12) imply that the polynomial c_2 satisfies $c_2(1) = i$, $c_2(\omega) = 0$, and $c_2(\omega^3) = c_1(\omega^3)$.

(c) Using the result of (b) and the BCH bound, show that C_m is a $[2n+1, 2n-3m, 5]$ code.

The code $\mathcal{P}(2)$ is a union of cosets of C_m, where each coset is determined by the value of $\sum_{x \in X} x$. We consider the coset $C_m(i)$ for which this value is the field element ω^i. Let $f(\xi)$ be the idempotent polynomial defined in Example 14.3, so $f(\omega) = 1$ and $f(\omega^i) = 0$ for all powers i not belonging to the same cyclotomic coset as 1.

(d) Show that $C_m(i)$ is obtained by adding the word $(\xi^i, 0, \xi^i f(\xi))$ to all the words of C_m.

REMARK. This exercise describes the original definition of Preparata codes. It has the advantage that the structure of the linear subcode C_m is easy to understand and often useful in further treatment of these codes. However, finding the minimum distance using this description is quite difficult.

9. Recall the 4-dimensional vector space W/Z over F_3 used to define the (5,5,2) partial geometry in (7.15): W is the set of 6-tuples over F_3 with sum 0, Z the span of the all-1 vector.

Show that the 1-dimensional subspaces of W/Z (the points of PG(3,3)) can be divided into three types, with coset representatives of the form $0^4 1^1 2^1$, $0^2 1^2 2^2$, $0^3 1^3$ respectively. (Here $0^4 1^1 2^1$ means a vector with four 0s, one 1 and one 2.) Let S_i be the set of points of type i. Show that $|S_i| = 15, 15, 10$ for $i = 1, 2, 3$ respectively. (These types are orbits of the symmetric group S_6.)

There is a natural bijection between the points and the hyperplanes of PG(3,3) (with this description): the hyperplane corresponding to the point $[w + Z]$ is

$$\{[v + Z] \ : \ \langle v, w \rangle = 0\}.$$

Hence the hyperplanes can also be divided into three types.

Verify the following table, whose (i, j) entry is the number of points of type j incident with a hyperplane of type i.

	$0^4 1^1 2^1$	$0^2 1^2 2^2$	$0^3 1^3$
$0^4 1^1 2^1$	6	3	4
$0^2 1^2 2^2$	3	6	4
$0^3 1^3$	6	6	1

Thus each of the sets S_1 and S_2 meets every hyperplane in 3 or 6 points, while the set S_3 meets every hyperplane in 1 or 4 points. Verify that the code and strongly regular graph derived from S_1 are those of Example (16.32). What is the set S_3?

10. This exercise outlines direct constructions of strongly regular graphs from a set $S \subseteq \mathrm{PG}(n-1, q)$, with the property that any hyperplane meets S in a or b points.

(a) Let p be a point of $\mathrm{PG}(n-1,q)$. Show that the number of pairs (r,s) of distinct points of $S \setminus \{p\}$ such that p,r,s are collinear depends only on whether or not $p \in S$. [Consider the case where $p \notin S$; the other case is similar. Let x, resp. y, be the number of hyperplanes containing p and meeting S in a, resp. b points. Show that

$$x + y = \frac{q^{n-1}-1}{q-1},$$

$$xa + yb = |S| \cdot \frac{q^{n-2}-1}{q-1},$$

ansd deduce that x and y are determined. Now show that the required number β of pairs satisfies

$$xa(a-1) + yb(b-1) = \beta \cdot \frac{q^{n-2}-1}{q-1} + (|S|(|S|-1)-\beta) \cdot \frac{q^{n-3}-1}{q-1}.]$$

(b) Construct a graph Γ_1 as follows. The vertex set is the underlying vector space $V = \mathbb{F}_q^n$; vertices $\mathbf{x}$ and $\mathbf{y}$ are adjacent if and only if $[\mathbf{x}-\mathbf{y}] \in S$. Use (a) to show that the graph Γ_1 is strongly regular.

(c) Show that the set S^* of hyperplanes meeting S in a points has the property that any point lies in a^* or b^* members of S^*, for some numbers a^* and b^*. In other words, S^* satisfies the same conditions as S, with a^* and b^* replacing a and b.

(d) Now construct Γ_2 as follows. The vertex set is the dual space of V. Two vertices $\mathbf{f}$ and $\mathbf{g}$ are joined if and only if the hyperplane $\ker(\mathbf{f}-\mathbf{g})$ meets S in a points. Prove that Γ_2 is strongly regular.

(e) Which of the two strongly regular graphs in the text ((16.22) and (16.23)) is Γ_1, and which is Γ_2?

(f) Now suppose that σ is a polarity of $\mathrm{PG}(n-1,q)$, and S a set of points with the property that $|p^\sigma \cap S|$ depends only on whether or not $p \in S$. Show that the strongly regular graphs Γ_1 and Γ_2 constructed above are isomorphic. (This situation occurs in Example (16.28), where S is a quadric and σ the associated polarity.)

17. Association schemes

Association schemes form a class of combinatorial structures including many types we have met, such as strongly regular graphs, square, quasi-symmetric, and tight designs, regular two-graphs, and systems of linked square designs. Their definition leads naturally to the use of algebraic methods in their study. So important and natural are they that closely related objects have been defined several times: in statistics by R. C. Bose and his co-workers (for example, Bose and Shimamoto (1952), Bose and Mesner (1959)), in group theory by D. G. Higman (1971a, b), and also in group theory by B. Weisfeiler (1976) and others (though the last two formulations were slightly more general). The most important contribution to the subject was the thesis of P. Delsarte (1973).

Since the last edition of our book, two specialized accounts have been published, by Bannai and Ito (1984), and by Brouwer, Cohen and Neumaier (1989). We do not attempt the same level of coverage.

To motivate the definition: we have seen several times that a strongly regular graph and its complement carry essentially the same information, so we should think of them together as a set on which three relations are defined: equality, adjacency, and non-adjacency. The definition ensures that the vector space spanned by the matrices of these three relations (viz. I, A, and $J-I-A$, where $A = A(\Gamma)$), is closed under multiplication. Similarly, a square design, its complement, and their duals are described by a set (the set of points and blocks) with four relations: equality; same type (both points or both blocks); incidence; and non-incidence. Note that, in all cases, equality is one of the basic relations: we take this for granted, and assume that the other relations are irreflexive and symmetric, and so can be represented by graphs.

(17.1) DEFINITION. An *association scheme* consists of a set X together with a partition of the set of 2-element subsets of X into d non-empty classes $\Gamma_1, \ldots, \Gamma_d$, satisfying the two conditions

(a) given $x \in X$, the number $n_i(x)$ of points $y \in X$ with $\{x, y\} \in \Gamma_i$ depends only on i, not on x (so we write this number as n_i);

(b) given $x, y \in X$ with $\{x, y\} \in \Gamma_k$, the number $p_{ij}^k(x, y)$ of points $z \in X$ with $\{x, z\} \in \Gamma_i$ and $\{z, y\} \in \Gamma_j$ depends only on i, j, k, not on x and y (so we write this number as p_{ij}^k).

Points x and y are called i^{th} *associates* if $\{x, y\} \in \Gamma_i$.

It is convenient to take a set of n 'colours' $c_1, \ldots, c_n$, and colour an edge of the complete graph on X with colour c_i if it belongs to Γ_i. Then Γ_i is the c_i-coloured subgraph. Condition (a) asserts that each monochrome subgraph is regular. Condition (b) asserts that the number of triangles with prescribed colouring on a given base depends only on the colouring, not on the chosen base. The reader should check the following easy result.

(17.2) Proposition. *In an association scheme with two classes, the two monochrome subgraphs are strongly regular. Conversely, a complementary pair of strongly regular graphs forms a two-class association scheme.* □

It will be convenient to enrich the picture slightly by adding a loop at each point, and assigning a new colour c_0 to each loop (or regarding the loops as edges of a highly degenerate graph Γ_0). Then condition (b) holds even if i, j, k are allowed to take the value 0 (and some of x, y, z to be equal); we take $p_{ii}^0 = n_i$ (where $n_0 = 1$), $p_{0i}^i = 1$, and $p_{ij}^k = 0$ if one of i, j, k is zero and the other two are unequal.

(17.3) EXAMPLE. Two important classes of association schemes are the Hamming and Johnson schemes. The point set of the *Hamming scheme* $H(n,q)$ are all ordered n-tuples of elements from an alphabet A of size q; two n-tuples are i^{th} associates if their Hamming distance is i, for $i = 1, \ldots, d$. In the *Johnson scheme* $J(v,k)$, with $k \leq \frac{1}{2}v$, the point set is the set of all k-element subsets of a fixed v-set; two k-sets are i^{th} associates if their symmetric difference has cardinality $2i$. As we will see, Delsarte used the Hamming and Johnson schemes as settings for classical coding theory and design theory respectively.

The following construction shows why association schemes and their generalizations are of interest to group theorists. It extends our remark about rank 3 permutation groups in Chapter 2.

Let G be a group of permutations of a set X. Suppose that any two distinct points of X are interchanged by some element of G. (Such a group is called *generously transitive*; clearly this property is a strengthening of transitivity.) The set of 2-element subsets of X falls into orbits $\Gamma_1, \ldots, \Gamma_d$ under the action of G. This partition is an association scheme. (To see that condition (b) holds, suppose that $\{x, y\}$ and $\{x', y'\}$ belong to the same orbit Γ_k. Then some element of G carries $\{x, y\}$ to $\{x', y'\}$, and so carries $\{z : \{x, z\} \in \Gamma_i, \{z, y\} \in \Gamma_j\}$ to $\{z : \{x', z\} \in \Gamma_i, \{z, y'\} \in \Gamma_j\}$. It follows that $p_{ij}^k(x, y) = p_{ij}^k(x', y')$.

REMARK. Various generalizations of the notion of an association scheme have been proposed. Usually, the basic relations are taken to be sets of ordered (rather than unordered) pairs; it is assumed that, instead of all the relations Γ_i being symmetric, they satisfy $\Gamma_i^* = \Gamma_{i^*}$ for some i^*, where Γ_i^* is the *converse* of Γ_i, that is,

$$\Gamma_i^* = \{(y,x) : (x,y) \in \Gamma_i\}.$$

If the definition (17.1) is taken over *verbatim*, we obtain what Higman (1971a,b) calls a *homogeneous coherent configuration*. In particular, any transitive permutation group gives rise to such a configuration, by a slight modification of the construction in the preceding paragraph. However, this is too general for our purposes. Delsarte, in his thesis (1973), admits configurations satisfying these axioms and the additional condition that

$$p_{ij}^k = p_{ji}^k$$

for all i, j, k. For such structures, almost all the results of this chapter remain true, with some modifications (mainly, replacing the real by the complex field in the algebraic arguments). Where necessary, we refer to these structures as *association schemes in the sense of Delsarte.*

(17.4) DEFINITION. A d-class association scheme is called *metric* if, for $i = 1, \ldots, d$, a pair $\{x, y\}$ belongs to Γ_i if and only if the distance from x to y in the graph Γ_1 is equal to i. (This definition depends on the ordering of the association classes $\Gamma_1, \ldots, \Gamma_d$. Strictly speaking, a scheme is metric if there is an ordering of the association classes for which the above condition holds.)

Note, however, that once Γ_1 is given, the rest of the ordering is determined. A connected graph Γ of diameter d is called *distance-regular* if, setting $\Gamma_i = \{\{x,y\} : d(x,y) = i\}$ for $i = 1, \ldots, d$ (where $d(x,y)$ is the distance in the graph Γ, we obtain an association scheme. The following result is true.

(17.5) Proposition. *Let Γ be a connected graph of diameter d. For $i = 1, \ldots, d$, let $\Gamma_i(x)$ denote the set of vertices at distance i from x. Then Γ is distance-regular if and only if there exist positive integers c_i, a_i, b_i ($i = 0, \ldots, d$) such that, for all x, y with $y \in \Gamma_i(x)$,*

$$|\Gamma_j(x) \cap \Gamma(y)| = c_i, a_i, \text{ or } b_i$$

according as $j = i-1, i, i+1$.

A d-class association scheme is metric if and only if
(a) $p_{ij}^k = 0$ unless $|i-j| \le k \le i+j$;
(b) $p_{i1}^{i+1} > 0$ for $i = 0, \ldots, d-1$. □

The distance-regular graphs which 'generate' the Hamming and Johnson association schemes, in the sense described above, are called the Hamming and Johnson graphs, and are denoted by the same symbols.

(17.6) REMARK. It can happen that more than one of the association graphs in a scheme is distance-regular; so the scheme can be metric in more than one way. For example, if $v = 2k+1$, the graph Γ_k in the Johnson scheme $J(v,k)$ is distance-regular. (This is the so-called 'odd graph', whose vertices are the k-subsets of a $2k+1$-set, two vertices adjacent if they are disjoint. For $k = 2$ it is the Petersen graph.) For a description of the ways in which this phenomenon can occur, see Bannai and Bannai (1980), or Bannai and Ito (1984), §III.4.

Now a number of identities connect the parameters of an association scheme; for example,

$$\begin{aligned} p_{ij}^k &= p_{ji}^k, \\ n_k p_{ij}^k &= n_i p_{kj}^i, \\ \sum_{j=0}^d p_{ij}^k &= n_i, \\ \sum_{t=0}^d p_{hi}^t p_{tj}^k &= \sum_{s=0}^d p_{hs}^k p_{ij}^s. \end{aligned} \tag{17.7}$$

All these equations can be proved by counting arguments. For example, the last equation comes from counting the number of paths with colour sequence (c_h, c_i, c_j) from x to y, where $\{x, y\} \in \Gamma_k$, in two different ways (Fig. 17.1).

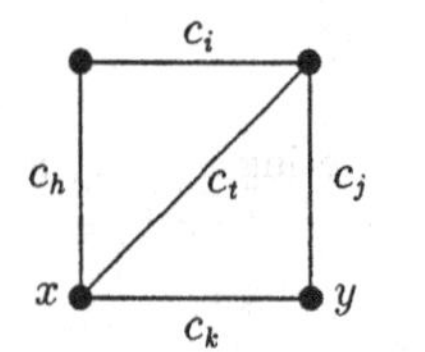

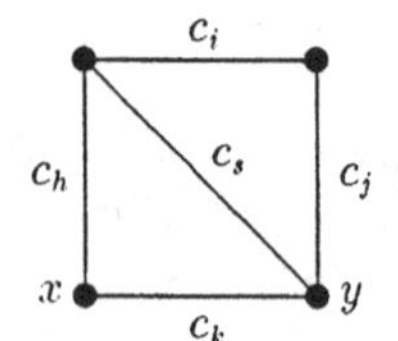

Fig. 17.1. Counting paths

Let A_i be the adjacency matrix of Γ_i for $i = 1, \ldots, d$, and $A_0 = I$. Then the (x, y) entry of $A_i A_j$ is the number of paths with colour sequence (c_i, c_j) from x to y, and so is p_{ij}^k if $\{x, y\} \in \Gamma_k$. Thus, we have the equation:

$$A_i A_j = \sum_{k=0}^d p_{ij}^k A_k. \tag{17.8}$$

So the symmetric matrices $A_0, \ldots, A_d$ commute pairwise (by the first equation of (17.7)), and their d-dimensional span (over $\mathbb{R}$) is closed under multiplication, and so is a commutative algebra $\mathcal{A}$.

The equations of (17.7) all have algebraic interpretations. For example, the fourth equation is the associative law

$$(A_h A_i) A_j = A_h (A_i A_j).$$

If the scheme is metric, then the equations (17.8) read

$$A_iA_1 = b_{i-1}A_{i-1} + a_iA_i + c_{i+1}A_{i+1}. \quad (17.9)$$

Thus, $A_2, A_3, \dots, A_d$ can be expressed inductively as polynomials in A_1 of degrees $2, 3, \dots, d$; so A_1 generates $\mathcal{A}$.

(17.10) DEFINITION. The algebra $\mathcal{A}$ is called the *Bose–Mesner algebra* of the association scheme. In the case of a scheme arising from a generously transitive permutation group, it is sometimes called the *centralizer algebra* of the group (since it consists of all those matrices which commute with all the permutation matrices in the group).

Just in the case of strongly regular graphs, the eigenvalues of the A_i and their multiplicities can be calculated from the parameters, and provide powerful non-existence criteria. Note that since $A_0, \dots, A_d$ are commuting symmetric matrices, they can be simultaneously diagonalized. The calculation of their spectra is made easier by the next result.

(17.11) Theorem. *With the notation introduced earlier, let P_i be the matrix with (j,k) entry p_{ij}^k, for $k = 0, \dots, d$. Then $P_0, \dots, P_d$ generate an algebra isomorphic to $\mathcal{A}$ (under the map $A_i \mapsto P_i$, $i = 0, \dots, d$.)*

PROOF. The fact that $P_iP_j = \sum_{k=0}^d p_{ij}^k P_k$ follows from the identities (17.5): the (s,t) entries of the two sides of the equation are $\sum_{k=0}^d p_{is}^k p_{jk}^t$ and $\sum_{k=0}^d p_{ij}^k p_{ks}^t$. □

REMARK. This isomorphism can be explained by noting that it is the 'regular representation' of $\mathcal{A}$; in other words, the image of $A \in \mathcal{A}$ under the isomorphism is the matrix representing left multiplication by A, relative to the basis $A_0, \dots, A_d$ of $\mathcal{A}$. (This is clear from the definition if $A = A_i$ for some i, and extends by linearity to all of $\mathcal{A}$.)

The matrices P_i are the *intersection matrices* or *reduced adjacency matrices* of the scheme. They are $(d+1) \times (d+1)$ matrices, and so are normally very much smaller than the adjacency matrices of the scheme. The eigenvalues of A_i are the same as those of P_i, and the latter can be much more easily computed.

The matrices $A_0, \dots, A_d$ are commuting symmetric matrices, and so have a simultaneous basis of eigenvectors (Cohn (1974), p. 203). By (17.11), the same is true of $P_0, \dots, P_d$. Let $\{\mathbf{u}_0, \dots, \mathbf{u}_d\}$ be such a basis, and let $\mathbf{u}_jP_i = p_i(j)\mathbf{u}_j$ for each i, j, so that $p_i(j)$ is the jth eigenvalue of P_i (or of A_i, by (17.11)). The function $A_i \mapsto p_i(j)$, for $i = 0, \dots, d$, extends linearly to a *character* of the algebra $\mathcal{A}$, denoted by π_j. The *multiplicity* of this character is the dimension of the simultaneous eigenspace. Just as for strongly regular graphs, it is possible to compute these multiplicities; the *integrality condition*, asserting that these numbers are non-negative integers, is a powerful non-existence criterion for association schemes.

(17.12) Theorem. *Let π_j be a character of the algebra $\mathcal{A}$. Let $\mathbf{u}_j$ and $\mathbf{v}_j^\top$ be left and right eigenvectors of the matrices P_i with eigenvalue $\pi_j(A_i) = p_i(j)$ for $j = 0, \ldots, d$, each normalized to have first entry 1. Then the multiplicity m_j of π_j is given by*

$$m_j = \frac{|X|}{\langle \mathbf{u}_j, \mathbf{v}_j \rangle}.$$

PROOF. This argument contains several ideas of wider significance! Let $n = |X|$, and let n_i, p_{ij}^k be the parameters of the scheme. Let P and Q be the matrices with i, j entries $p_j(i)$ and $q_j(i) = m_j p_i(j)/n_i$ respectively. The matrices P and Q are called the *first* and *second eigenmatrices* of the scheme. We have

$$\begin{aligned} p_i(k)p_j(k)\mathbf{v}_k &= P_i P_j \mathbf{v}_k \\ &= \sum_{l=0}^{d} p_{ij}^l P_l \mathbf{v}_k \\ &= \sum_{l=0}^{d} p_{ij}^l p_l(k) \mathbf{v}_k, \end{aligned}$$

so $\sum_{l=0}^{d} p_{ij}^l p_l(k) = p_i(k)p_j(k)$. It follows that the vector

$$(p_0(k), \ldots, p_d(k))^\top$$

is an eigenvector of P_i with eigenvalue $p_i(k)$. By definition, this vector is equal to $\mathbf{v}_k^\top$.

The trace of $A_{i_1} A_{i_2} \ldots A_{i_k}$ is equal to the number of closed paths with colour sequence $i_1, i_2, \ldots, i_k$. In particular, we have

$$\begin{aligned} \mathrm{Trace}(A_i) &= n\delta_{i0}, \\ \mathrm{Trace}(A_i A_j) &= n n_i \delta_{ij}, \qquad (17.13) \\ \mathrm{Trace}(A_i A_j A_k) &= n n_k p_{ij}^k, \end{aligned}$$

where δ_{ij} is the Kronecker delta. that is.

$$\delta_{ij} = \begin{cases} 1 & \text{if } i = j, \\ 0 & \text{if } i \neq j. \end{cases}$$

The second equation in (17.13), in terms of eigenvalues, yields

$$\begin{aligned} n n_i \delta_{ij} &= \sum_{k=0}^{d} m_k p_i(k) p_j(k) \\ &= n_i \sum_{k=0}^{d} p_j(k) q_k(j), \end{aligned}$$

by definition of $q_k(j)$. Thus, the eigenmatrices P and Q satisfy

$$QP = nI. \qquad (17.14)$$

Hence $PQ = nI$, giving

$$n = m_k \sum_{i=0}^{d} \frac{(p_i(k))^2}{n_i},$$

or $m_k = n/\left(\sum_{i=0}^{d}(p_i(k)^2)/n_i\right)$.

It also follows from (17.14) that

$$\left(\frac{p_0(k)}{n_0}, \ldots, \frac{p_d(k)}{n_d}\right)$$

is a left eigenvector of P_i with eigenvalue $p_i(k)$, and hence is equal to $\mathbf{u}_k$. So we have

$$m_k = \frac{n}{\langle \mathbf{u}_k, \mathbf{v}_k \rangle},$$

as required. □

(17.15) DEFINITION. Assume the notation defined above. The association scheme is said to be *P-polynomial* if there are polynomials $f_0, f_1, \ldots, f_d$, where f_j has degree j, so that $p_i(j) = f_j(p_1(j))$ for all i, j.

Note that this is equivalent to the statement that $P_i = f_i(P_1)$, and hence to $A_i = f_i(A_1)$, by (17.11). In other words, we have the following.

(17.16) Proposition. *An association scheme is P-polynomial if and only if it is metric.* □

(17.17) EXAMPLE. The Hamming scheme $H(d, q)$ is metric, and hence it is P-polynomial. The eigenvalues of the basis matrices are given by the Krawtchouk polynomials which we met (for $q = 2$) in Chapter 14. Specifically, $p_i(j) = K_i(j)$, where

$$K_i(x) = \sum_{j=0}^{i}(-1)^j(q-1)^{i-j}\binom{x}{j}\binom{n-x}{i-j}.$$

Then the polynomials f_i in (17.13) differ from these only by a simple affine transformation, viz.

$$f_i(x) = K_i(d - \frac{d+x}{q}).$$

(The last assertion follows because $K_1(x) = (q-1)d - qx$, so that $x = K_1(d - \frac{d+x}{q})$.)

We refer to Exercise 4, or Bannai and Ito (1984), §III.2, for the proof.

Let E_k denote the orthogonal projection of the $\mathbf{R}^n$ onto the eigenspace affording the character π_k. By definition, we have the spectral decomposition

$$A_i = \sum_{k=0}^{d} p_i(k) E_k.$$

By (17.14), it follows that

$$E_i = n^{-1} \sum_{k=0}^{d} q_i(k) A_k.$$

The matrices $E_0, \dots, E_d$ thus form another basis for the Bose–Mesner algebra. In this basis, the multiplication has the simple form

$$E_i E_j = \delta_{ij} E_i;$$

in other words, the E_i are orthogonal idempotents.

There is a second multiplication defined on the Bose–Mesner algebra $\mathcal{A}$, namely *pointwise* or *Hadamard multiplication*: if $A = (a_{ij})$, $B = (b_{ij})$, then $A \circ B = C = (c_{ij})$, where

$$c_{ij} = a_{ij} b_{ij}.$$

We met this operation in Chapter 2, but we will consider it in greater detail here. The basis matrices $A_0, \dots, A_d$ satisfy

$$A_i \circ A_j = \delta_{ij} A_i,$$

in other words, they form a basis of orthogonal idempotents for $\mathcal{A}, \circ$. The Hadamard product of two Es lies in $\mathcal{A}$, and so is expressible in terms of the Es:

$$E_i \circ E_j = \sum_{k=0}^{d} q_{ij}^k E_k, \tag{17.18}$$

where

$$q_i(m) q_j(m) = \sum_{k=0}^{d} q_{ij}^k q_k(m).$$

Equation (17.18) shows that the numbers q_{ij}^k, for $k = 0, \dots, d$, are the eigenvalues of $E_i \circ E_j$. By (2.25), we draw the following conclusion, known as the *Krein bound*.

(17.19) Theorem. *For all i, j, k, we have $q_{ij}^k \geq 0$.* □

The parameters q_{ij}^k satisfy various relations similar to (17.7). We do not give details.

For any polynomial $f(x)$ and matrix A, we will use $(\circ f)(A)$ for the result of evaluating f at A using Hadamard multiplication. For example, if $f(x) = x^2 + 3x + 1$, then $(\circ f)(A) = A \circ A + 3A + J$. (Note that J is the identity for $\circ$.)

(17.20) DEFINITION. An association scheme is said to be *Q-polynomial* (with respect to the ordering $E_0, \dots, E_d$ of the ordinary idempotents) if there exist polynomials $q_0, \dots, q_d$, where q_i has degree i for $0 \leq i \leq d$, such that

$$E_i = (\circ q_i)(E_1)$$

for $i = 0, \dots, d$.

An equivalent condition is that
(a) $q_{ij}^k = 0$ unless $|i-j| \leq k \leq i+j$;
(b) $q_{i1}^{i+1} > 0$ for $i = 0, \ldots, d-1$.

Only a small proportion of all association schemes have either the P-polynomial or the Q-polynomial property. Nevertheless, of those which do, (and especially those with a fairly large number of classes), very many have both of these properties. These include the schemes of greatest importance for design theory and coding theory, the Hamming and Johnson schemes, as well as others such as schemes of alternating bilinear forms which are relevant to the material in Chapter 12. Consequently, these schemes have received the most attention.

D. A. Leonard (1984) proved the following remarkable result.

(17.21) Theorem. *All the parameters of a P- and Q-polynomial association scheme can be expressed in terms of five quantities.* □

The explicit expressions for the parameters are very complicated. The full statement of Leonard's theorem in Bannai and Ito (1984) covers eleven pages, the bulk of which consists of formulae for the parameters, subdivided into nine cases. We refer to Bannai and Ito (1984), §III.5, or Brouwer, Cohen and Neumaier (1989), §8.1, for details.

The significance of Leonard's theorem is that it brings the complete classification of P- and Q-polynomial association schemes within reach. Bannai and Ito (1984) regard this as the main problem of algebraic combinatorics. Such a program requires a large number of classifications of specific classes of association schemes by their parameters. See Exercise 3 for a simple example of such a classification, and Bannai and Ito (1984) for comments on the general problem. We remark that particular classification theorems require analysis of the clique structure of the graphs in question, in the same spirit as Bose (1963) (as we discussed in Chapter 7).

The proof of (17.12) above suggests that a formal 'duality' holds between the first and second eigenmatrices of an association scheme. Under this duality, the intersection numbers and the Krein parameters (the latter suitably normalized, by multiplying by $|X|$) correspond, as do the basis matrices and the minimal idempotents, matrix multiplication and Hadamard product, valencies and multiplicities, the P-polynomial and Q-polynomial conditions, etc. We will see below that various coding-theoretic and design-theoretic properties of subsets of an association scheme also correspond under this duality. We shall say that one scheme is the *formal dual* of another if the first eigenmatrix of the first scheme coincides with the second eigenmatrix of the second scheme (up to ordering of rows and columns).

Delsarte (1973) found a situation in which there is an actual duality between association schemes, which implies the formal duality just defined.

Let A be an abelian group which is a group of automorphisms of the association scheme $(X, \{\Gamma_0, \ldots, \Gamma_d\})$, and acts transitively on X. In this situation, A is sharply transitive on X, and we can set up a bijection between X and A as follows: select $x \in X$; then $y \in X$ corresponds to $g \in A$ if g maps x to y. Under this correspondence, the given action of A on X translates into the action of A on itself by right translation. We now assume that this has been done, and take $X = A$. Moreover, we write A additively, so that the identity element is denoted by 0.

For $0 \leq i \leq d$, let $C_i = \{a \in A : \{0, a\} \in \Gamma_i\}$. Now the sets C_i determine the scheme: a and b are i^{th} associates if and only if $b - a \in C_i$.

Now let A^* be the group of characters of A. (Note that A^* is an abelian group with the same order as A; but the group operation is naturally written as multiplication, since it is given by $(\phi \cdot \chi)(a) = \phi(a) \cdot \chi(a)$.) For $\phi, \chi \in A^*$, write $\phi \sim \chi$ if $\phi(C_i) = \chi(C_i)$ for $0 \leq i \leq d$, where $\phi(S) = \sum_{s \in S} \phi(s)$. This is obviously an equivalence relation on A^*; let $C_0^* = \{1\}, \ldots, C_e^*$ be its equivalence classes. Delsarte proved the following.

(17.22) Theorem. *With the above notation, $d = e$; moreover, if we call $\phi, \chi \in A^*$ i^{th} associates whenever $\phi \cdot \chi^{-1} \in C_i^*$, then we obtain an association scheme, which is a formal dual to the original scheme.* □

REMARK. There are close connections with the theory of Schur rings: see Tamaschke (1963).

The Hamming schemes admit abelian transitive automorphism groups. If we give the alphabet A the structure of an abelian group of order q, then the Cartesian power A^n acts on the points of $H(n, q)$. (If A is a field, this is just the group of translations of the vector space.) The dual of $H(n, q)$, in the sense of (17.22), turns out to be isomorphic to $H(n, q)$ again. (See Exercise 12.) In particular, the first and second eigenmatrices of $H(n, q)$ are equal.

We come now to the central material of Delsarte's thesis, in which association schemes appear as a natural setting for coding and design theory, and these two theories are related by formal duality.

Let $(X, \{\Gamma_0, \ldots, \Gamma_d\})$ be an association scheme. We assume fixed but arbitrary orderings of the basis matrices and of the minimal idempotents in the Bose–Mesner algebra, subject to the condition that $A_0 = I$ and $E_0 = (1/n)J$. Let P and Q be the eigenmatrices of the scheme.

Let C be a subset of X, with $|C| > 1$. The *distribution vector* of C is the vector $\mathbf{a} = (a_0, \ldots, a_d)$, where

$$a_i = \frac{1}{|C|} |\{(x, y) \in C \times C : x \text{ and } y \text{ are } i^{\text{th}} \text{ associates}\}|.$$

Its *Q-transform* is the vector $\mathbf{b} = \mathbf{a}Q$.

(17.23) Theorem. *The distribution vector and its Q-transform are non-negative vectors.*

PROOF. The non-negativity of $\mathbf{a}$ is clear. We have

$$|C|b_k = |C|\sum_{i=0}^{d} a_i Q_k(i),$$

which is the sum of all the entries of the submatrix indexed by C of the matrix $\sum_{i=0}^{d} A_i Q_k(i) = E_k$. Now E_k is positive semi-definite (having eigenvalues 0 and 1); so, if $\mathbf{c}$ is the characteristic function of C, we have

$$|C|b_k = \mathbf{c}E_k\mathbf{c}^{\top} \geq 0,$$

as required. □

This result generalizes (14.22), and forms the basis of the most general version of the linear programming bound.

Now let M be a non-empty subset of $\{1,\ldots,d\}$. We say that the set C is an *M-code* if $a_i = 0$ for all $i \in M$; and C is an *M-design* if $b_i = 0$ for all $i \in M$. Thus, M-codes and M-designs are configurations extremal with respect to specified subsets of the inequalities (17.23). We investigate what these concepts mean in specific cases.

Clearly, C is an M-code if and only if no two elements of C are i^{th} associates for any $i \in M$; in other words, C is a coclique in the graph obtained by amalgamating the Γ_i for $i \in M$. Suppose, in particular, that the scheme is P-polynomial (that is, metric), with the basis matrices ordered in the natural way, and that $M = \{1,\ldots,t\}$ for some $t < d$. Then an M-code (or *t-code*, as we say in this particular case), is a set C of points satisfying

$$\forall_{x,y\in C}\,[x \neq y \Rightarrow d(x,y) > t],$$

where d is the distance in the graph Γ_1: this is an obvious generalization of the definition of a code with minimum distance greater than t, as given in Chapter 9.

Many concepts of coding theory extend to this situation. For example, we have the *sphere-packing bound*:

(17.24) Theorem. *A $2e$-code C in a P-polynomial scheme satisfies*

$$|C| \leq |X| / \left(\sum_{i=0}^{e} n_i\right).$$

□

This, of course, holds because the balls of radius e with centres at the points of C are pairwise disjoint, and each such ball contains $\sum_{i=0}^{e} n_i$ points. A code attaining the bound is called *perfect*. Perfect codes in various distance-regular graphs have been examined, and sometimes classified; see Biggs (1973), (1974), Cameron *et al.* (1976), for example.

The interpretation of the concept of M-design is much less clear. We restrict attention to the case where the association scheme is Q-polynomial (with the idempotents ordered in the natural way), and $M = \{1, \ldots, t\}$ (in which case we refer to a t-design). Even here, we have no general interpretation; it is necessary to work out the meaning of the condition in each case. We give the most important example.

(17.25) Theorem. *A t-design in the Johnson scheme $J(k, v)$ is the same thing as a t-design in the usual sense.*

PROOF. Let X be a v-set. We use the matrices N_{ji} of the proof of (1.51) : for $i \leq j$, N_{ij} is the $\binom{v}{j} \times \binom{v}{i}$ matrix whose rows and columns are indexed by the j-subsets and i-subsets of X respectively, having (J, I) entry 1 if $I \subseteq J$, 0 otherwise. Set $N_i = N_{ki}$.

We leave it to the reader to show that, for any $i \leq k$, the space spanned by the matrices $E_0, \ldots, E_i$ is the same as the space spanned by $N_0 N_0^\top, \ldots, N_i N_i^\top$.

Recall that $|C| b_i$ is equal to the sum of all entries of the matrix E_i lying in rows or columns indexed by C. Assume that C is a t-design in $J(v, k)$, and let $\mathbf{c}$ be the characteristic function of C. Then, for $i = 1, \ldots, t$, we have

$$\mathbf{c} E_i \mathbf{c}^\top = 0.$$

Moreover, we have $\mathbf{1} E_i \mathbf{1}^\top = 0$ for $i = 1, \ldots, t$ (since $E_0 E_i = O$ and $E_0 = (1/n)J$); and

$$(\mathbf{c} - \alpha \mathbf{1}) E_0 (\mathbf{c} - \alpha \mathbf{1})^\top = 0,$$

where $\alpha = |C| / \binom{v}{k}$.

By the preceding paragraph, it follows that

$$(\mathbf{c} - \alpha \mathbf{1}) N_t N_t^\top (\mathbf{c} - \alpha \mathbf{1})^\top = 0,$$

whence $\mathbf{c} N_t = \alpha \mathbf{1} N_t = \lambda \mathbf{1}$ for some number λ.

But entries in the vector $\mathbf{c} N_t$ count the number of k-sets in C which contain a given t-subset. So (X, C) is a t-design.

Conversely, a t-design is an i-design for all $i \leq t$; then the above argument reverses to show that $\mathbf{c} E_i \mathbf{c}^\top = 0$ for $i = 1, \ldots, t$, whence C is a t-design in the Johnson scheme. □

Thus, the inequalities (17.23) give conditions satisfied by the Q-transform of a subset of the Johnson scheme, so that the extremal sets are precisely the t-designs. These have been worked out by S. Hobart (1989), who gave an elementary (and remarkably short!) proof of the following result. We have outlined her argument in Exercise 10.

(17.26) Theorem. *Let* a *be the distribution vector of a subset S of $J(v,k)$, of cardinality b. Then*

$$\sum_{i=0}^{k} \binom{i}{t} a_i \geq b \binom{k}{t}^2 \Big/ \binom{v}{t},$$

with inequality if and only if S is the block set of a t-design. □

We mention the next result without proof. Recall from Chapter 14 that a set C of n-tuples over an alphabet A is an *orthogonal array of strength t* if there is a positive constant λ such that, for any choice of t distinct coordinates $i_1, \ldots, i_t$, and any choice of elements $a_1, \ldots, a_t \in A$, then number of words $\mathbf{w} \in C$ for which $w_{i_k} = a_k$ for $k = 1, \ldots, t$, is precisely λ.

(17.27) Theorem. *A non-empty subset of the Hamming scheme $H(n,q)$ is a t-design in the scheme if and only if it is an orthogonal array of strength t.* □

Delsarte proved the following result, dual to (17.23).

(17.28) Theorem. *Let C be a $2e$-design in a Q-polynomial association scheme with multiplicities $m_0, \ldots, m_d$. Then*

$$|C| \geq \sum_{i=0}^{e} m_i.$$

□

(17.29) EXAMPLE. Consider the Johnson scheme $J(v,k)$. We saw that, in the notation of (17.25), the matrix $N_e N_e^\top$, with rank $\binom{v}{e}$, is a positive combination of the idempotents $E_0, \ldots, E_e$. Thus,

$$\sum_{i=0}^{e} m_i = \binom{v}{e},$$

and (17.28) reduces exactly to the theorem of Ray-Chaudhuri and Wilson (1.50).

(17.30) EXAMPLE. In the Hamming scheme $H(n,q)$, the multiplicities are the same as the degrees, viz.

$$m_i = n_i = (q-1)^i \binom{n}{i}.$$

Thus, we obtain *Rao's inequality* (Rao (1947)): an orthogonal array C of strength $2e$ in $H(n,q)$ satisfies

$$|C| \geq \sum_{i=0}^{e} (q-1)^i \binom{n}{i}.$$

Tight orthogonal arrays of strength 4 have been determined by Noda (1979).

Further refinements of these results involve Delsarte's 'four fundamental parameters' defined in Chapter 14. These are generalized to association schemes as follows.

(17.31) DEFINITION. (a) Let C be a non-empty subset of an association scheme, having distribution vector $\mathbf{a}$. The *minimum distance* of C is the integer

$$d = \min\{i \geq 1 : a_i \neq 0\},$$

and the *number of non-zero distances* of C is the integer

$$s = |\{i \geq 1 : a_i \neq 0\}|.$$

(b) Let C be a non-empty subset of an association scheme, having distribution vector $\mathbf{a}$ with Q-transform $\mathbf{b}$. The *dual distance* of C is the integer

$$d' = \min\{i \geq 1 : b_i \neq 0\},$$

and the *external distance* of C is the integer

$$r = |\{i \geq 1 : b_i \neq 0\}|.$$

REMARKS. 1. The minimum distance and the dual distance are only meaningful if a fixed ordering of the basis matrices, resp. the idempotents, is given. In practice, we only consider minimum distance in P-polynomial schemes, and dual distance in Q-polynomial schemes. These parameters are one greater than the largest t for which C is a t-code, resp. a t-design, in the sense previously defined.

2. The terms *degree* and *strength* are also used for s and $d'-1$, respectively.

In the next theorem, the valencies of the association scheme are n_i, and the multiplicities m_i, as usual.

(17.32) Theorem. *(a) Let C be a non-empty subset of a P-polynomial association scheme, having minimum distance d and external distance r. Put $e = \lfloor \frac{1}{2}(d-1) \rfloor$. Then*

$$\sum_{i=0}^{e} n_i \leq |X|/|C| \leq \sum_{i=0}^{r} n_i.$$

In particular, $e \leq r$. If one of these bounds is attained, then all three are attained (and C is a perfect code).

(b) Let C be a non-empty subset of a Q-polynomial association scheme, having dual distance d' and number of non-zero distances s. Put $e = \lfloor \frac{1}{2}(d'-1) \rfloor$. Then

$$\sum_{i=0}^{e} m_i \leq |C| \leq \sum_{i=0}^{s} m_i.$$

In particular, $e \leq s$. If one of these bounds is attained, then all three are attained (and C is a tight design). □

(17.33) EXAMPLE. Let C be a linear code in $H(n,q)$, and $C^\perp$ its dual (where q is a prime power). Then the distribution vectors $\mathbf{a}$ and $\mathbf{a}'$ of C and $C^\perp$ satisfy

$$|C|\mathbf{a}' = \mathbf{a}Q, \qquad |C^\perp|\mathbf{a} = \mathbf{a}'Q.$$

(Recall that $P = Q$ for the Hamming schemes, and $PQ = |X|I$, where $|X| = |C| \cdot |C^\perp|$.) This assertion is equivalent to the MacWilliams identities (9.15) (see (14.23)). In particular, the minimum distance and external distance of $C^\perp$ are equal to the dual distance and number of distances, respectively, of C. More can be said. For example, if the induced structure on C is an association scheme, then the space $X/C^\perp$ — which can be identified with the group of characters of C — carries in a natural way an association scheme, which is the dual of C. (See Exercise 6 for a particular example of this situation.)

Sometimes, pairs of non-linear codes have the properties that their distribution vectors are Q-transforms of each other, up to a factor. We saw this phenomenon for the Preparata and Kerdock codes. No satisfactory explanation is known.

One more result will suffice to illustrate Delsarte's work. If C is a subset of the point set of an association scheme $(X, \{\Gamma_1, \ldots, \Gamma_d\})$, then C carries an induced structure given by the graphs $\Gamma_i|C$ which are non-empty (that is, for which $a_i \neq 0$ — there are s of these). When does it happen that this induced structure is an association scheme with s classes? Delsarte showed the following.

(17.34) Theorem. *Let C be a subset of a Q-polynomial association scheme, having number of distances s and strength t (that is, dual distance $t+1$). Suppose that $t \geq 2s-2$. Then the induced structure on C is an association scheme with s classes, having multiplicities $m_0, \ldots, m_s - 1$, and $|C| - \sum_{i=0}^{s-1} m_i$.*

(17.35) REMARK. The theorem applies when X is a Johnson scheme, and $s = 2$, $t \geq 2$, in which case C is the set of blocks of a quasi-symmetric 2-design. In this case, (17.34) reduces to (5.3).

Exercises

1. Show that, for n even, the Hamming scheme $H(n,2)$ is metric with respect to the ordering $\Gamma_{n-1}, \Gamma_2, \Gamma_{n-3}, \Gamma_4, \ldots, \Gamma_3, \Gamma_{n-2}, \Gamma_1, \Gamma_n$.

2. Let G be a finite group, with conjugacy classes $C_0 = \{1\}, \ldots, C_d$. Define relations $\Gamma_0, \ldots, \Gamma_d$ on G by

$$\Gamma_i = \{(x,y) \in G \times G | x^{-1}y \in C_i\}.$$

(a) Show that these relations are invariant under left and right multiplication by elements of G.
(b) Show that $(G, \{\Gamma_0, \ldots, \Gamma_d\})$ is an association scheme in the sense of Delsarte.
(c) Show that the second eigenmatrix of this association scheme is the character table of the group G.

3. Let Γ be a distance-regular graph with parameters $c_i = i$, $a_i = 0$, $b_i = d - i$ for $0 \le i \le d$. Let x be a fixed vertex, and $\Gamma_i(x)$ the set of vertices at distance i from x.
(a) Show that the map $y \mapsto \Gamma_1(x) \cap \Gamma_{i-1}(y)$ is a bijection between $\Gamma_i(x)$ and the set of i-element subsets of the d-set $\Gamma_1(x)$, for $0 \le i \le d$.
(b) Hence show that Γ is a Hamming graph $H(d, 2)$.

4. Show that, for fixed d and q, the Krawtchouk polynomials $K_i(x)$ satisfy the recurrence

$$(q-1)(d-i+1)K_{i-1}(x) + i(q-2)K_i(x) + (i+1)K_{i+1}(x) = K_1(x)K_i(x),$$

where $K_1(x) = (q-1)d - qx$, and (by convention) $K_{-1}(x) = K_{d+1}(x) = 0$. Deduce that, for $x = 0, 1, \ldots, d$, the vector

$$(K_0(x), K_1(x), \ldots, K_d(x))^\top$$

is an eigenvector for the intersection matrix P_1 of the Hamming scheme $H(d, q)$, with eigenvalue $K_1(x)$. Deduce (17.17).

5. Let A be the additive group of $\mathbb{F}_q$, and let d be a divisor of $q-1$ such that, if q is odd, then $(q-1)/d$ is even. Let K be the multiplicative group of dth roots of unity in $\mathbb{F}_q^\times$. (By assumption, $-1 \in K$.) Let $C_1, \ldots, C_d$ be the cosets of K, and $C_0 = \{0\}$. Prove that the structure constructed from $C_0, \ldots, C_d$ by the recipe preceding (17.22) is an association scheme. (These association schemes are called *cyclotomic schemes* $C(q, d)$.) Prove that
(a) for $q \equiv 1 \pmod 4$, the Paley graph $P(q)$ is a class in $C(q, 2)$;
(b) the Clebsch graph is a class in $C(16, 3)$;
(c) it is possible to choose three classes of $C(81, 8)$ whose union is a strongly regular graph with parameters (81, 30, 9, 12) (and is the point graph of the Van Lint–Schrijver partial geometry of Example (7.20)).
Show that cyclotomic schemes are formally self-dual.

REMARK. If we consider association schemes in the sense of Delsarte, it is not necessary to require that $(q-1)/d$ is even in this construction.

6. Let Γ_1 be the strongly regular graph of (11.12), with parameters (243, 22, 1, 2). Let Γ_2 be the strongly regular graph of Chapter 11, Exercise 9, with parameters (243, 132, 81, 60). Show that the association schemes defined by Γ_1 and Γ_2 are formal duals of each other. Are they actual duals (in the sense of (17.22))?

7. (a) Verify that the two strongly regular graphs associated with a set of points in projective space having just two cardinalities of hyperplane sections ((16.22) and (16.23), or Exercise 10 of Chapter 16) give rise to association schemes which are formal duals of each other.

(b) Prove that a 2-class association scheme is formally self-dual if and only if the two strongly regular graphs which define the scheme are of pseudo- or negative Latin square type or conference graphs. [Hint: use the eigenvalues k, r, s of one of the graphs as parameters. In fact, the conclusion follows from the assumption that $\{k, l\} = \{f, g\}$.]

8. What does it mean to say that a set of k-sets is an e-error correcting code in the Johnson scheme? Show that a necessary condition for a perfect 1-error correcting code in $J(v, k)$ is that $1 + k(v - k)$ divides $\binom{v}{k}$. Prove that, if $k = 3$ and $v \geq 6$, then a perfect 1-error correcting code in $J(v, k)$ can exist only for $v = 6$, and find the unique example with $v = 6$.

9. Show that a perfect 1-error-correcting code in the odd graph of (17.6) (the graph Γ_k in the Johnson scheme $J(2k + 1, k)$) is the set of blocks of a $(k - 1)$-$(2k + 1, k, 1)$ design, and conversely.

10. (a) Give an example of a 3-design with exactly 3 distances between blocks, whose induced structure in the Johnson scheme is not an association scheme.

(b) Show that, if a 3-design has three distances, and has the further property that the complement of any block is a block, then the induced structure is an association scheme.

11. Let S be a set of b points of the Johnson scheme $J(v, k)$ (that is, k-subsets of a v-set X). For distinct $p_1, \ldots, p_t \in X$, let $\lambda_{\{p_1,\ldots,p_t\}}$ be the number of members of S which contain $p_1, \ldots, p_t$. Let $(a_0, \ldots, a_k)$ be the distribution vector of S.

By counting choices of $\{p_1, \ldots, p_t\} \subseteq X$ and j members of S containing $p_1, \ldots, p_t$, for $j = 1, 2$, show that

$$\sum \lambda_{\{p_1,\ldots,p_t\}} = b\binom{k}{t},$$

$$\sum (\lambda_{\{p_1,\ldots,p_t\}})^2 = b\sum_{i=0}^{k}\binom{i}{t}a_i.$$

Now use the variance trick to show Hobart's Theorem (17.26).

12. Use Theorem (17.22) to show that the Hamming scheme $H(d, 2)$ is self-dual. [HINT: Let A be the translation group. The characters of A are bijective with words; $\mathbf{w} \leftrightarrow \phi_{\mathbf{w}}$, where $\phi_{\mathbf{w}}(\mathbf{v}) = (-1)^{(\mathbf{v},\mathbf{w})}$. Use (14.21) to show that the equivalence class of the relation $\sim$ defined before (17.22) are the sets

$$C_i^* = \{\phi_{\mathbf{w}} : \mathrm{wt}(\mathbf{w}) = i\}$$

for $0 \leq i \leq d$.]

13. (a) Let X be the set of points and blocks of a square 2-(v,k,λ) design, carrying the three relations

- same type (both points, or both blocks);
- incident;
- different type and non-incident.

Show that X is a 3-class association scheme.

(b) Let Γ be a graph with vertex set V. Let $X = \{v^+, v^- : v \in V\}$, and define a graph Γ^* with vertex set X in which v^δ and w^ϵ are adjacent if and only if $\delta\epsilon = b_{vw}$, where $B = (b_{vw})$ is the $(0,-,+)$ adjacency matrix of Γ introduced in Chapter 4. Show that switching Γ does not change the isomorphism type of Γ^*, so that it is an invariant of the two-graph Δ derived from Γ. Show also that, if Δ is a non-trivial regular two-graph, then Γ^* is a distance-regular graph of diameter 3 (and so defines a 3-class association scheme).

(c) Show that a formal dual of an association scheme of type (a) above (if it exists) is of type (b), and *vice versa.*

(d) Let D be a subset of an abelian group A of order v, with $|D| = k$. Suppose that

- for all $a \in A$, $a \neq 0$, the number of pairs $(d_1, d_2) \in D^2$ with $a = d_1 - d_2$ is precisely λ;
- for all $d \in D$, $-d \in D$.

(Such a set is called a *difference set with multiplier* -1.) Show that

$$(A, \{D + a : a \in A\})$$

is a square 2-(v,k,λ) design. Show further that the association scheme derived from this design as in (a) admits a transitive abelian group of automorphisms isomorphic to $\mathbf{Z}_2 \times A$. Deduce that there exists a regular two-graph on v points whose eigenvalues have multiplicities k and $v-k$.

(e) Take $A = \mathbf{Z}_4 \times \mathbf{Z}_4$, $D = \{(0,x),(x,0) : x \in \mathbf{Z}_4, x \neq 0\}$. Verify that D is a difference set with multiplier -1. What is the corresponding regular two-graph?

REMARK. Parts (c) and (d) are due to M. C. Whelan (1989), who also used a difference set with multiplier -1 and parameters $(4000, 775, 150)$ due to R. L. McFarland (1973) to construct a new regular two-graph on 4000 points.

References

R. W. Ahrens and G. Szekeres (1969), On a combinatorial generalization of 27 lines associated with a cubic surface, *J. Austral. Math. Soc*, **10**, 485–492.

E. F. Assmus, Jr. & J. H. van Lint (1979), Ovals in projective designs, *J. Combinatorial Theory* (A), **27**, 307–324.

E. F. Assmus, Jr. & H. F. Mattson, Jr. (1969), New 5-designs, *J. Combinatorial Theory* (A), **6**, 122–151.

E. F. Assmus, Jr. & H. F. Mattson, Jr. (1971), Algebraic theory of codes II, *Report AFCRL-0013, Appl. Research Lab. of Sylvania Electronic Systems*, Bedford, Mass.

E. F. Assmus, Jr. & H. F. Mattson, Jr. (1972a), On weights in quadratic residue codes, *Discrete Math.*, **3**, 1–20.

E. F. Assmus, Jr. & H. F. Mattson, Jr. (1972b), Contractions of self-orthogonal codes, *Discrete Math.*, **3**, 21–32.

E. F. Assmus, Jr., H. F. Mattson, Jr. & Marcia Guza (1974), Self-orthogonal Steiner systems and projective planes, *Math. Z.*, **138**, 89–96.

E. F. Assmus, Jr., J. A. Mezzaroba & C. J. Salwach (1977), Planes and biplanes, *Higher Combinatorics* (M. Aigner, ed.), 205–212, D. Reidel, Dordrecht.

E. F. Assmus, Jr. & H. F. Sachar (1977), Ovals from the point of view of coding theory, *Higher Combinatorics* (M. Aigner, ed.), 213–216, D. Reidel, Dordrecht.

E. F. Assmus, Jr. & C. J. Salwach (1979), The $(16, 6, 2)$ designs, *Internat. J. Math. Math. Sci.*, **2**, 261–281.

A. Baartmans & M. S. Shrikhande (1982), Designs with no three mutually disjoint blocks, *Discrete Math.*, **40**, 129–139.

B. Bagchi (1988), No extendable biplane of order 9, *J. Combinatorial Theory* (A), **49**, 1–12.

B. Bagchi (1991), Corrigendum: No extendable biplane of order 9, *J. Combinatorial Theory* (A), **57**, 162.

R. D. Baker, J. H. van Lint, and R. M. Wilson (1983), On the Preparata and Goethals codes, *IEEE Trans. Inform. Theory* **29**, 342–345.

E. E. Bannai (1972), Maximal subgroups of low rank of finite symmetric and alternating groups, *J. Fac. Sci. Univ. Tokyo*, **18**, 475–486.

E. Bannai (1977), On tight designs, *Quart. J. Math. Oxford*, (2) **28**, 433–448.

E. Bannai and E. Bannai (1980), How many P-polynomial structures can an association scheme have? *Europ. J. Combinatorics*, **1**, 289–298.

E. Bannai & T. Ito (1973), On finite Moore graphs, *J. Fac. Sci. Univ. Tokyo*, **20**, 191–208.

E. Bannai and T. Ito (1984), *Algebraic Combinatorics I: Association Schemes*, Benjamin, New York.

A. Barlotti (1955), Un'estensione del teorema di Segre-Kustaanheimo, *Boll. Un. Mat. Ital*, **10**, 498–506.

L. W. Beineke & R. J. Wilson (1978), *Selected Topics in Graph Theory*, Academic Press, London.

L. W. Beineke & R. J. Wilson (1983), *Selected Topics in Graph Theory II*, Academic Press, London.

E. R. Berlekamp, J. H. van Lint & J. J. Seidel (1973), A strongly regular graph derived from the perfect ternary Golay code, *A Survey of Combinatorial Theory* (ed. J. N. Srivastava *et al.*), 25–30, North-Holland, Amsterdam.

E. R. Berlekamp, F. J. MacWilliams, & N. J. A. Sloane (1972), Gleason's theorem on self-dual codes, *IEEE Trans. Inform. Theory*, **18**, 409–414.

M. R. Best (1982), A contribution to the nonexistence of perfect codes, Ph. D. Thesis, University of Amsterdam.

K. N. Bhattacharya (1944), On a new symmetrical balanced incomplete block design, *Bull. Calcutta Math. Soc.*, **36**, 91–96.

N. L. Biggs (1971), *Finite Groups of Automorphisms*, *London Math. Soc. Lecture Notes* **6**, Cambridge Univ. Press, Cambridge.

N. L. Biggs (1973), Perfect codes in graphs, *J. Combinatorial Theory* (B), **15**, 289–296.

N. L. Biggs (1974), Perfect codes and distance-transitive graphs, *Combinatorics* (T, P, McDonough & V. C. Mavron, eds.), 1–8, *London Math. Soc. Lecture Notes* **13**, Cambridge Univ. Press, Cambridge.

A. Blokhuis (1984), On subsets of GF(q^2) with square differences, *Proc. Kon. Ned. Akad. v. Wetensch.* (A), **87**, 369–372.

A. Blokhuis & A. E. Brouwer (1984), Uniqueness of a Zara graph on 126 points and non-existence of a completely regular two-graph on 288 points, *Papers dedicated to J. J. Seidel* (ed. P. J. de Doelder, J. de Graaf & J. H. van Lint), 6–19, Report 84-WSK-03, Eindhoven Univ. Technology.

A. Blokhuis & H. Wilbrink (1989), Characterization theorems for Zara graphs, *Europ. J. Combinatorics*, **10** (1989), 57–68.

R. C. Bose (1942), A note on the resolvability of balanced incomplete block designs, *Sankhya*, **6**, 105–110.

R. C. Bose (1963), Strongly regular graphs, partial geometries, and partially balanced designs, *Pacific J. Math.*, **13**, 389–419.

R. C. Bose & D. M. Mesner (1959), On linear associative algebras corresponding to association schemes of partially balanced designs, *Ann. Math. Statist.*, **30**, 21–38.

R. C. Bose & T. Shimamoto (1952), Classification and analysis of partially balanced incomplete block design with two associate classes, *J. Amer. Statist. Assoc.*, **47**, 151–184.

R. C. Bose, S. S. Shrikhande & N. M. Singhi (1977), Edge-regular multigraphs and partial geometric designs, *Teorie Combinatorie* (Tomo I), Accad. Naz. Lincei, Roma.

A. Bremner (1979), A diophantine equation arising from tight 4-designs, *Osaka J. Math.*, **167**, 353–356.

A. E. Brouwer, A. M. Cohen & A. Neumaier (1989), *Distance-Regular Graphs*, Springer-Verlag, Berlin.

R. H. Bruck (1963), Finite nets II: Uniqueness and embedding, *Pacific J. Math.*, **13**, 421–457.

R. H. Bruck & H. J. Ryser (1949), The nonexistence of certain finite projective planes, *Canad. J. Math.*, **1**, 88–93.

A. Bruen & J. C. Fisher (1973), Blocking sets, k-arcs and nets of order ten, *Advances in Math.*, **10**, 317–320.

J. M. J. Buczak (1980), *Some Topics in Group Theory*, Ph. D. Thesis, Oxford University.

F. Buekenhout (1969), Une caractérisation des espaces affines basée sur la notion de droite, *Math. Z.*, **111**, 367–371.

F. Buekenhout and E. E. Shult (1974), On the foundations of polar geometry, *Geometriae Dedicata*, **3**, 155–170.

F. C. Bussemaker & J. J. Seidel (1970), Symmetric Hadamard matrices of order 36, *Ann. N. Y. Acad. Sci.*, **175**, 66–79.

F. C. Bussemaker, D. M. Cvetković & J. J. Seidel (1978), Graphs related to exceptional root systems, *Combinatorics* (ed. A. Hajnal & V. T. Sós), pp. 185–191, North-Holland, Amsterdam.

A. R. Calderbank (1987), The application of invariant theory to the existence of quasi-symmetric designs, *J. Combinatorial Theory* (A), **44**, 94–109.

A. R. Calderbank (1988a), Inequalities for quasi-symmetric designs, *J. Combinatorial Theory* (A), **46**, 53–64.

A. R. Calderbank (1988b), Geometric invariants for quasi-symmetric designs, *J. Combinatorial Theory* (A), **47**, 101–110.

A. R. Calderbank & W. M. Kantor (1986), The geometry of two-weight codes, *Bull. London Math. Soc.*, **18**, 97–122.

P. J. Cameron (1973a), Extending symmetric designs, *J. Combinatorial Theory* (A), **14**, 215–220.

P. J. Cameron (1973b), Biplanes, *Math. Z.*, **131**, 85–101.

P. J. Cameron (1978), On doubly transitive permutation groups of degree prime squared plus one, *J. Austral. Math. Soc.* (A), **26**, 317–318.

P. J. Cameron (1980a), 6-transitive graphs, *J. Combinatorial Theory* (B), **28**, 168–179.

P. J. Cameron (1980b), A note on generalized line graphs, *J. Graph Theory*, **4**, 243–245.

P. J. Cameron (1991), Several 2-(46, 6, 3) designs, *Discrete Math.*, **87**, 89–90.

P. J. Cameron (to appear), Quasi-symmetric designs admitting a spread, *Combinatorics '88, proc. Ravello Conf.*, Mediterranean Press, Roma.

P. J. Cameron, J.-M. Goethals & J. J. Seidel (1978a), Strongly regular graphs having strongly regular subconstituents, *J. Algebra*, **55**, 257–280.

P. J. Cameron, J.-M. Goethals & J. J. Seidel (1978b), The Krein condition, spherical designs, Norton algebras and permutation groups, *Proc. Kon. Nederl. Akad. Wetensch.* (A), **81**, 196–206 (= *Indag. Math.*, **40**, 196–206).

P. J. Cameron, J.-M. Goethals, J. J. Seidel & E. E. Shult (1976), Line graphs, root systems, and elliptic geometry, *J. Algebra*, **43**, 305–327.

P. J. Cameron, D. R. Hughes & A. Pasini (1990), Extended generalized quadrangles, *Geometriae Dedicata*, **35**, 193–228.

P. J. Cameron & J. H. van Lint (1982), On the partial geometry $pg(6,6,2)$, *J. Combinatorial Theory* (A), **32**, 252–255.

P. J. Cameron & C. E. Praeger (to appear), Block-transitive designs, I: Point-imprimitive designs.

P. J. Cameron & J. J. Seidel (1973), Quadratic forms over GF(2), *Proc. Kon. Nederl. Akad. Wetensch.* (A), **76**, 1–8 (= *Indag. Math.*, **35**, 1–8).

P. J. Cameron, J. A. Thas & S. E. Payne (1976), Polarities of generalized hexagons and perfect codes, *Geometriae Dedicata*, **5**, 525–528.

P. Camion (1975), Global quadratic abelian codes, *Information Theory* (G. Longo, ed.), *CISM Courses & Lectures* **219**, Springer-Verlag, Vienna.

Chang Li-Chien (1959), The uniqueness and non-uniqueness of the triangular association schemes, *Sci. Record Peking Math.* (New Ser.), **3**, 604–613.

Chang Li-Chien (1960), Association schemes of partially balanced designs with parameters $v = 28$. $n_1 = 12$, and $p^2_{11} = 4$, *Sci. Record Peking Math.* (New Ser.), **4**, 12–18.

S. Chowla, P. Erdős & E. G. Straus (1960), On the maximal number of pairwise orthogonal Latin squares of a given order, *Canadian J. Math.*, **12**, 204–208.

S. Chowla & H. J. Ryser (1950), Combinatorial problems, *Canad. J. Math.*, **2**, 93–99.

P. M. Cohn (1974), *Algebra*, Vol. 1, Wiley, London.

P. M. Cohn (1977), *Algebra*, Vol. 2, Wiley, London.

J. H. Conway (1969), A group of order 8,315,553,613,086,720,000, *Bull. London Math. Soc.*, **1**, 79–88.

J. H. Conway (1976), The Miracle Octad Generator, *Topics in Group Theory and Computation, Proc. Summer School, Galway 1973*, 62–68, Academic Press, London.

J. H. Conway, R. T. Curtis, S. P. Norton, R. A. Parker, & R. A. Wilson (1985), **ATLAS** *of Finite Groups*, Oxford University Press, Oxford.

R. T. Curtis (1976), A new combinatorial approach to M_{24}, *Math. Proc. Cambridge Philos. Soc.*, **79**, 25–42.

R. M. Damerell (1973), On Moore graphs, *Proc. Cambridge Philos. Soc.*, **74**, 227–236.

I. Debroey & J. A. Thas (1978), On semipartial geometries, *J. Combinatorial Theory* (A), **25**, 242–250.

P. Delsarte (1971), Majority logic decodable codes derived from finite inversive planes, *Inform. Control*, **18**, 319–325.

P. Delsarte (1972), Weights of linear codes and strongly regular normed spaces, *Discrete Math.*, **3**, 47–64.

P. Delsarte (1973), An algebraic approach to the association schemes of coding theory, *Philips Res. Reports Suppl.* **10**.

P. Delsarte & J.-M. Goethals (1975), Alternating bilinear forms over GF(q), *J. Combinatorial Theory* (A), **19**, 26–50.

P. Delsarte, J.-M. Goethals & J. J. Seidel (1971), Orthogonal matrices with zero diagonal, II, *Canad. J. Math.*, **23**, 816–832.

P. Delsarte, J.-M. Goethals & J. J. Seidel (1977), Spherical codes and designs, *Geometriae Dedicata*, **6**, 363–388.

P. Dembowski (1963), Inversive planes of even order, *Bull. Amer. Math. Soc.*, **69**, 850–854.

P. Dembowski (1964a), Ein Kennzeichnung der endlichen affinen Räume, *Arch. Math.*, **15**, 146–154.

P. Dembowski (1964b), Möbiusebenen gerader Ordnung, *Math. Ann.*, **157**, 179–205.

P. Dembowski (1968), *Finite Geometries*, Springer-Verlag, Berlin.

P. Dembowski & A. Wagner (1960), Some characterizations of finite projective spaces, *Arch. Math.*, **11**, 465–469.

R. H. F. Denniston (1969), Some maximal arcs in finite projective planes, *J. Combinatorial Theory*, **6**, 317–319.

R. H. F. Denniston (1976), Some new 5-designs, *Bull. London Math. Soc.*, **8**, 263–267.

R. H. F. Denniston (1979), On biplanes with 56 points, *Ars Combinatoria*, **7**, 221–222.

M. Deza (1973), Une propriété extrémale des plans projectifs finis dans une classe de codes équidistants, *Discrete Math.*, **6**, 353–358.

L. E. Dickson (1958), *Linear groups with an exposition of the Galois field theory*, Dover, New York.

J. F. Dillon (1974), On Pall partitions for quadratic forms, Ph. D. Thesis Univ. of Maryland.

J. Doyen, X. Hubaut & M. Vandensaval (1978), Ranks of incidence matrices of Steiner triple systems, *Math. Z.*, **163**, 251–259.

L. M. H. E. Driessen, G. H. M. Frederix, & J. H. van Lint (1976), Linear codes suppported by Steiner triple systems, *Ars Combinatoria*, **1**, 33–42.

R. H. Dye (1977), Partitions and their stabilizers for line complexes and quadrics, *Ann. Mat.* (4), **114**, 173–194.

H. Enomoto, N. Ito & R. Noda (1979), Tight 4-designs, *Osaka J. Math.*, **16**, 39–43.

P. Erdős, A. Rényi, & V. T. Sós (1966), On a problem in graph theory, *Studies Math. Hungar.*, **1**, 215–235.

W. Feit & G. Higman (1964), The nonexistence of certain generalized polygons, *J. Algebra*, **1**, 434–446.

D. Foulser (1969), Solvable primitive permutation groups of low rank, *Trans. Amer. Math. Soc.*, **143**, 1–54

F. R. Gantmacher (1959), *Applications of the Theory of Matrices*, Interscience, New York.

A. D. Gardiner (1976), Homogeneous graphs, *J. Combinatorial Theory* (B), **20**, 94–102.

A. Gewirtz (1969), Graphs of maximal even girth, *Canad. J. Math.*, **21**, 915–934.

A. M. Gleason (1978), Weight polynomials of self-dual codes and the MacWilliams identities, *Actes Congrès Internat. Math.* (Vol. 3), 211–215.

J.-M. Goethals (1971), On the Golay perfect binary code, *J. Combinatorial Theory* (A), **11**, 178–186.

J.-M. Goethals (1973), Some combinatorial aspects of coding theory, *A Survey of Combinatorial Theory* (J. N. Srivastava *et al.*, eds.), Chapter 17, North-Holland, Amsterdam.

J.-M. Goethals & J. J. Seidel (1967), Orthogonal matrices with zero diagonal, *Canad. J. Math.*, **19**, 1001–1010.

J.-M. Goethals & J. J. Seidel (1970), Strongly regular graphs derived from combinatorial designs, *Canad. J. Math.*, **22**, 597–614.

J.-M. Goethals & S. Snover (1972), Nearly perfect binary codes, *Discrete Math.*, **2**, 65–88.

M. J. E. Golay (1949), Notes on digital coding, *Proc. IEEE* **37**, 657.

D. Y. Goldberg (1986), Reconstructing the ternary Golay code, *J. Combinatorial Theory* **42**, 269–279.

B. H. Gross (1974), Intersection triangles and block intersection numbers for Steiner systems, *Math. Z.*, **139**, 87–104.

J. Hadamard (1893), Résolution d'une question relative aux déterminants, *Bull. Sci. Math.* (2), **17**, 240–246.

W. H. Haemers (1981), A new partial geometry constructed from the Hoffman-Singleton graph, *Finite Geometries and Designs* (Proc. Second Isle of Thorns Conf. 1980), 119–127, *London Math. Soc. Lecture Notes*, **49**, Cambridge University Press.

J. I. Hall (1982), Classifying copolar spaces and graphs, *Quart. J. Math. Oxford* (2), **33**, 421–429.

J. I. Hall, A. J. E. M. Janssen, A. W. J. Kolen & J. H. van Lint (1977), Equidistant codes with distance 12, *Discrete Math.*, **17**, 71–83.

M. Hall, Jr. (1960), Automorphisms of Steiner triple systems, *IBM J. Res. Develop.*, **4**, 460–472.

M. Hall, Jr. (1971), Affine generalized quadrilaterals, *Studies Pure Math.*, 113–116.

M. Hall, Jr. (1986), *Combinatorial Theory*, Wiley, New York.

M. Hall, Jr. & W. S. Connor (1953), An embedding theorem for balanced incomplete block designs, *Canad. J. Math.*, **6**, 35–41.

M. Hall, Jr., R. Lane & D. Wales (1970), Designs derived from permutation groups, *J. Combinatorial Theory* (A), **8**, 12–22.

N. Hamada (1973), On the p-rank of the incidence matrix of a balanced or partially balanced incomplete block design and its application to error-correcting codes, *Hiroshima Math. J.*, **3**, 153–226.

G. H. Hardy & E. M. Wright (1981), *An Introduction to the Theory of Numbers*, Oxford University Press, Oxford.

D. G. Higman (1971a), *Combinatorial Considerations about Permutation Groups*, Math. Institute, Oxford.

D. G. Higman (1971b), Partial geometries, generalized quadrangles, and strongly regular graphs, *Atti del Conv. Geometria, Combinatoria e sue Applicazioni*, Perugia.

D. G. Higman (1973), Remark on Shult's graph extension theorem, *Finite Groups '72* (T. Gagen, M. P. Hale, Jr., & E. E. Shult, eds.), North-Holland, Amsterdam.

D. G. Higman & C. C. Sims (1968), A simple group of order 44,352,000, *Math. Z.*, **105**, 110–113.

S. A. Hobart (1989), A characterization of t-designs in terms of the inner distribution, *Europ. J. Combinatorics*, **10**, 445–448.

A. J. Hoffman (1960), On the uniqueness of the triangular association scheme, *Ann. Math. Statist.*, **31**, 492–497.

A. J. Hoffman (1977), On graphs whose least eigenvalue exceeds $-1-\sqrt{2}$, *Linear Algebra Appl.*, **16**, 153–166.

A. J. Hoffman & D. K. Ray-Chaudhuri (unpublished), On a spectral characterization of regular line graphs.

A. J. Hoffman & R. R. Singleton (1960), On Moore graphs of diameters 2 and 3, *IBM J. Res. Develop.*, **4**, 497–504.

G. Hölz (1981), Construction of designs which contain a unital, *Arch. Math.*, **37**, 179–183.

Y. Hong (1984), On the nonexistence of unknown perfect 6- and 8-codes in Hamming schemes $H(n,q)$ with q arbitrary, *Osaka J. Math.* **21**, 687–700.

D. R. Hughes (1961), On t-designs and groups, *Amer. J. Math.*, **87**, 761–778.

D. R. Hughes & F. C. Piper (1973), *Projective Planes*, Springer-Verlag, Berlin.

D. R. Hughes & F. C. Piper (1985), *Design Theory*, Cambridge Univ. Press, Cambridge.

N. Ito (1975), Tight 4-designs, *Osaka J. Math.*, *12*, 493–522.

S. M. Johnson (1962), A new upper bound for error-correcting codes, *IEEE Trans. Inform. Theory*, **8**, 203–207.

W. Jónsson (1972), On the Mathieu groups M_{22}, M_{23}, M_{24} and the associated Steiner systems, *Math. Z.*, **125**, 193–214.

W. M. Kantor (1974), Dimension and embedding theorems for geometric lattices, *J. Combinatorial Theory* (A), **17**, 173–195.

W. M. Kantor (1975), Symmetric designs, symplectic groups and line ovals, *J. Algebra*, **33**, 43–58.

W. M. Kantor (1977), Moore geometries and rank 3 groups having $\mu = 1$, *Quart. J. Math. Oxford* (2), **28**, 309–328.

W. M. Kantor (1982), An exponential number of generalized Kerdock codes, *Info. and Control*, **53**, 74–80.

W. M. Kantor (1983), Spreads, translation planes and Kerdock sets I, II, *SIAM J. Alg. Disc. Math.*, **3**, 151–165 and 308–318.

W. M. Kantor & R. A. Liebler (1982), The rank 3 permutation representations of the finite classical groups, *Trans. Amer. Math. Soc.*, **271**, 1–71.

A. M. Kerdock (1972), A class of low-rate nonlinear codes, *Info. and Control*, **20**, 182–187.

T. P. Kirkman (1847), On a problem in combinations, *Cambridge and Dublin Math. J.*, **2**, 191–204.

T. P. Kirkman (1850), Query, *Lady's and Gentleman's Diary*, (1850), 48.

C. W. Lam, J. McKay, S. Swiercz & L. Thiel (1983), The nonexistence of ovals in a projective plane of order 10, *Discrete Math.* **45**, 319–321.

C. W. Lam, S. Swiercz & L. Thiel (1986), The nonexistence of codewords of weight 16 in a projective plane of order 10, *J. Combinatorial Theory* (A) **42**, 207–214.

C. W. Lam, S. Swiercz & L. Thiel (1989), The nonexistence of finite projective planes of order 10, *Canad. J. Math.* **41**, 1117–1123.

E. S. Lander (1981), Symmetric designs and self-dual codes, *J. London Math. Soc.* (2), **24**, 193–204.

E. S. Lander (1983), *Symmetric Designs: an Algebraic Approach, London Math. Soc. Lecture Notes*, **74**, Cambridge Univ. Press, Cambridge.

P. W. Lemmens & J. J. Seidel (1973), Equiangular lines, *J. Algebra*, **24**, 494–512.

D. A. Leonard (1984), Parameters of association schemes that are both P- and Q-polynomial, *J. Combinatorial Theory* (A), **36**, 355–363.

M. W. Liebeck (1987), The affine permutation groups of rank 3, *Proc. London Math. Soc.* (3), **54**, 477–516.

M. W. Liebeck & J. Saxl (1986), The finite primitive permutation groups of rank 3, *Bull. London Math. Soc.*, **18**, 165–172.

K. Lindström (1975), The nonexistence of unknown nearly perfect binary codes, *Ann. Univ. Turku* (Ser. AI), **169**.

J. H. van Lint (1969), Equidistant point sets, in *Combinatorics*, T. P. McDonough and V. C. Mavron, eds., LMS Lecture Notes **13**, Cambridge Univ. Press, 169–176.

J. H. van Lint (1971a), Nonexistence theorems for perfect error-correcting codes, in *Computers in Algebra and Number Theory* Vol. IV, SIAM-AMS Proceedings.

J. H. van Lint (1971b), *Coding Theory, Lecture Notes in Math.* **201**, Springer-Verlag, Berlin.

J. H. van Lint (1974a), Recent results on perfect codes and related topics, *Combinatorics* (Part 1) (M. Hall, Jr. & J. H. van Lint, eds.), *Math. Centre Tracts*, **55**, 158–178, Math. Centre, Amsterdam.

J. H. van Lint (1974b), Equidistant point sets, *Combinatorics* (T. P. McDonough & V. C. Mavron, eds.), *London Math. Soc. Lecture Notes*, **13**, Cambridge Univ. Press, Cambridge.

J. H. van Lint (1978), Non-embeddable quasi-residual designs, *Proc. Kon. Nederl. Akad. Wetensch.* (A), **81**, 269–275.

J. H. van Lint (1982), *Introduction to Coding Theory, Graduate Texts in Math.* **86**, Springer-Verlag, Berlin.

J. H. van Lint (1984) Partial geometries, *Proc. Int. Congr. Math.* (Warszawa 1983), Vol. 2, 1579–1589.

J. H. van Lint & F. J. MacWilliams (1978), Generalized quadratic residue codes, *IEEE Trans. Inform. Theory*, **24**, 730–737.

J. H. van Lint & A. Schrijver (1981), Constructions of strongly regular graphs, two-weight codes and partial geometries by finite fields, *Combinatorica*, **1**, 63–73.

J. H. van Lint & J. J. Seidel (1966), Equilateral point sets in elliptic geometry, *Proc. Kon. Nederl. Akad. Wetensch.*, **69**, 335–348 (= *Indag. Math.*, **28**, 335–348).

H. Lüneburg (1965), *Die Suzukigruppen und ihre Geometrien, Lecture Notes in Math.*, Springer-Verlag, Berlin.

H. Lüneburg (1969), *Transitive Erweiterungen endlicher Permutationsgruppen, Lecture Notes in Math.* **84**, Springer-Verlag, Berlin.

F. J. MacWilliams (1963), A theorem on the distribution of weights in a systematic code, *Bell Syst. Tech. J.* **42**, 79–94.

F. J. MacWilliams & N. J. A. Sloane (1977), *The Theory of Error-Correcting Codes*, North-Holland, Amsterdam.

F. J. MacWilliams, N. J. A. Sloane & J.-M. Goethals (1972), The MacWilliams identities for non-linear codes, *Bell System Tech. J.*, **51**, 803–819.

F. J. MacWilliams, N. J. A. Sloane & J. G. Thompson (1973), On the existence of a projective plane of order 10, *J. Combinatorial Theory* (A), *14*, 66–78.

S. S. Magliveras & D. W. Leavitt (1983), Simple six designs exist, *Proc. 14th Southeastern Conf. Combinatorics, Graph Theory, Computing*, 195–205, *Congr. Num.* **40**, Utilitas Math., Winnipeg.

C. L. Mallows & N. J. A. Sloane (1973), An upper bound for self-dual codes, *Inform. Control*, **22**, 188–200.

J. L. Massey (1963), *Threshold Decoding*, MIT Press, Cambridge, Mass.

É. Mathieu (1861), Mémoire sur l'étude des fonctions des plusieurs quantités, sur le manière de les former et sur les substitutions qui les laissent invariables, *J. Math Pures Appl. (Liouville)* (2), **6**, 241–323.

É. Mathieu (1873), Sur la fonction cinq fois transitive de 24 quantités, *J. Math. Pures Appl. (Liouville)* (2), **18**, 25–46.

R. Mathon (1975), 3-class association schemes, *Proc. Conf. Algebraic Aspects of Combinatorics* (Toronto 1975), 123–155, *Congr. Num.* **13**, Utilitas Math., Winnipeg.

R. L. McFarland (1973), A family of difference sets in non-cyclic groups, *J. Combinatorial Theory* (A), **15**, 1–10.

J. E. McLaughlin (1969), A simple group of order 898,128,000, *Theory of Finite Groups* (ed. R. Brauer & C.-H. Sah), 109–111, Benjamin, New York.

D. M. Mesner (1965), A note on the parameters of PBIB association schemes, *Ann. Math. Statist.*, **36**, 331–336.

W. H. Mills (1978), A new 5-design, *Ars Combinatoria*, **6**, 193–195.

A. Mostowski (1945), Axiom of choice for finite sets, *Fund. Math.*, **33**, 137–168.

D. E. Muller (1954), Application of Boolean algebra to switching circuit design and to error detection, *IEEE Trans. Computers* **3**, 6–12.

A. Neumaier (1979), Strongly regular graphs with smallest eigenvalue $-m$, *Arch. Math.*, **42**, 89–96.

A. Neumaier (1982a), Completely regular two-graphs, *Arch. Math.*, **38**, 378–384.

A. Neumaier (1982b), Regular sets and quasi-symmetric 2-designs, *Combinatorial Theory* (ed. D. Jungnickel & K. Vedder), 258–275, *Lecture Notes in Math.* **969**, Springer, Berlin.

R. Noda (1974), On homogeneous systems of linked symmetric designs, *Math. Z.*, **138**, 15–20.

R. Noda (1979), On orthogonal arrays of strength 4 achieving Rao's bound, *J. London Math. Soc.* (2), **19**, 385–390.

R. E. A. C. Paley (1933), On orthogonal matrices, *J. Math. Phys. Mass. Inst. Tech.*, **12**, 311–320.

G. Pasquier (1980), The Golay code obtained from an extended cyclic code over F_8, *Eur. J. Combinatorics* **1**, 369–370.

A. J. L. Paulus (1973), Conference matrices and graphs of order 26, *Report 73-WSK-06*, Techn. Univ., Eindhoven.

R. M. Pawale (1991), Quasi-symmetric 3-designs with triangle-free graph, *Geometriae Dedicata*, **37**, 205–210.

S. E. Payne & J. A. Thas (1985), *Finite Generalized Quadrangles*, Pitman, New York.

A. Ja. Petrenjuk (1968), *Math. Zametski*, **4**, 417–425.

V. Pless (1972), Symmetry codes over GF(3) and new 5-designs, *J. Combinatorial Theory* (A), **12**, 119–142.

V. Pless (1975), Symmetry codes and their invariant subcodes, *J. Combinatorial Theory* (A), **18**, 116–125.

F. P. Preparata (1968), A class of optimum nonlinear double-error-correcting codes, *Inform. Control*, **13**, 378–400.

C. R. Rao (1947), Factorial experiments derivable from combinatorial arrangements of arrays, *J. Roy. Statist. Soc.*, **9**, 128–139.

D. K. Ray-Chaudhuri & R. M. Wilson (1975), On t-designs, *Osaka J. Math.*, **12**, 737–744.

I. S. Reed (1954), A class of multiple-error-correcting codes and the decoding scheme, *IEEE Trans. Inform. Theory*, **4**, 38–49.

H. J. Ryser (1965), *Combinatorial Mathematics*, Wiley, New York.

H. Sachar (1973), Error-correcting codes associated with projective planes, Ph. D. Thesis, Lehigh Univ., Bethlehem, Pa.

C. J. Salwach & J. A. Mezzaroba (1979), The four known biplanes with $k = 11$, *Internat. J. Math. and Math. Sci.*, **2**, 251–260.

S. S. Sane and M. S. Shrikhande (to appear), *Quasi-symmetric Designs*.

J. Schreier & S. M. Ulam (1937), Über die Automorphismen der Permutationsgruppe der natürlichen Zahlenfolge, *Fund. Math.*, **28**, 258–260.

L. L. Scott (1977), Some properties of character products, *J. Algebra*, **45**, 259–265.

B. Segre (1954), Sulle ovali nei piani lineari finiti, *Atti Accad. Naz. Lincei Rendic.*, **17**, 141–142.

J. J. Seidel (1967), Strongly regular graphs of L_2 type and of triangular type, *Proc. Kon. Nederl. Akad. Wetensch.* (A), **70**, 188–196 (= *Indag. Math.*, **29**, 188–196).

J. J. Seidel (1968), Strongly regular graphs with $(-1, 1, 0)$ adjacency matrix having eigenvalue 3, *Linear Algebra Appl.*, **1**, 281–298.

J. J. Seidel (1969), Strongly regular graphs, *Recent Progress in Combinatorics* (W. T. Tutte, ed.), 185–197, Academic Press, New York.

J. J. Seidel (1973), On two-graphs and Shult's characterization of symplectic and orthogonal geometries over GF(2), *T.H. Report 73-WSK-02*, Techn. Univ., Eindhoven.

J. J. Seidel (1974), Graphs and two-graphs, *Proc. Fifth Southeastern Conf. Combinstorics, Graph Theory, Computing*, 125–143, *Congr. Num. X*, Utilitas Math. Publ. Co., Winnipeg.

J. J. Seidel (1977), A survey of two-graphs, *Teorie Combinatorie*, Accad. Naz. Lincei, Roma.

J. J. Seidel & D. E. Taylor (1978), Two-graphs, a second survey, *Algebraic Methods in Graph Theory*, Szeged.

N. V. Semakov, V. A. Zinovjev, & G. V. Zaitsef (1971), Uniformly packed codes, *Probl. Peredaci Inform.*, 38–50; English transl. *Problems of Information Transmission*, **7**, 30–39.

S. S. Shrikhande (1959), The uniqueness of the L_2 association scheme, *Ann. Math. Statist.*, **30**, 781–798.

E. E. Shult (1972a), The graph extension theorem, *Proc. Amer. Math. Soc.*, **33**, 278–284.

E. E. Shult (1972b), Characterizations of certain classes of graphs, *J. Combinatorial Theory* (B), **13**, 142–167.

C. C. Sims (1969), On the isomorphism of two groups of orders 44,352,000, *Theory of Finite Groups* (R. Brauer & C.-H. Sah, eds.), Benjamin, New York.

T. Skolem (1931), Über einige besondere Tripelsysteme mit Anwendung auf die Reproduktion gewisser Quadratsummen bei Multiplikation, *Norsk Mat. Tidskr.*, **13**, 41-51.

N. J. A. Sloane (1973), Is there a (72, 36) $d = 16$ self-dual code? *IEEE Trans. Inform. Theory*, **19**, 251.

M. S. Smith (1975), On rank 3 permutation groups, *J. Algebra*, **33**, 22–42.

D. W. Stanton (1986), t-designs in classical association schemes, *Graphs & Combinatorics*, **2**, 283–286.

J. J. Sylvester (1844), Elementary researches in the analysis of combinatorial aggregation, *Philos. Mag.*, **24**, 285–296.

O. Tamaschke (1963), Zur Theorie der Permutationsgruppen mit regulärer Untergruppe I, II, *Math. Z.*, **80**, 328–352, 443–465.

D. E. Taylor (1977), Regular 2-graphs, *Proc. London Math. Soc.* (3), **35**, 257–274.

L. Teirlinck (1987), Nontrivial t-designs without repeated blocks exist for all t, *Discrete Math.*, **65**, 301–311.

L. Teirlinck (1989), Locally trivial t-designs and t-designs without repeated blocks, *Discrete Math.*, **77**, 345–356.

J. A. Thas (1975), Some results concerning $\{(q+1)(n-1);\ n\}$-arcs and $\{(q+1)(n-1)+1;\ n\}$-arcs in finite projective planes of order q, *J. Combinatorial theory* (A), *19*, 228–232.

J. A. Thas (1976), Extensions of finite generalized quadrangles, *Combinatorica*, pp. 127–143, *Symp. Math.* **28**, Academic Press, London.

J. A. Thas (1977), Combinatorics of partial geometries and generalized quadrangles, *Higher Combinatorics* (M. Aigner, ed.), 183–199, D. Reidel, Dordrecht.

A. Tietäväinen (1973), On the nonexistence of perfect codes over finite fields, *SIAM J. Appl. Math.*, **24**, 88–96.

H. C. A. van Tilborg (1976), Uniformly packed codes, Thesis, Technological Univ. Eindhoven.

J. A. Todd (1966), A representation of the Mathieu group M_{24} as a collineation group, *Ann. di Mat. Pura ed Appl.* (Ser. 4), **71**, 199–238.

V. D. Tonchev (1986), Quasi-symmetric designs and self-dual codes, *Europ. J. Combinatorics*, **7**, 67–73.

R. J. Turyn (1974), Hadamard matrices, Baumert-Hall units, four-symbol sequences, pulse compression, and surface wave encodings, *J. Combinatorial theory* (A), **16**, 313–333.

O. Veblen & J. W. Young (1916), *Projective Geometry*, Ginn & Co., Boston.

H. N. Ward (1974), Quadratic residue codes and symplectic groups, *J. Algebra*, **29**, 150–171.

B. Weisfeiler (ed.) (1976), *On Construction and Identification of Graphs*, *Lecture Notes in Math.*, **558**, Springer-Verlag, Berlin.

M. C. Whelan (1989), *Imprimitive Association Schemes*, Ph. D. thesis, University of London.

H. Whitney (1932), Congruent graphs and the connectivity of graphs, *Amer. J. Math.*, **54**, 150–168.

H. Wielandt (1964), *Finite Permutation Groups*, Academic Press, New York.

R. M. Wilson (1973), The necessary conditions for t-designs are sufficient for something, *Utilitas Math.*, **4**, 207–215.

R. M. Wilson (1974a), Concerning the number of mutually orthogonal Latin squares, *Discrete Math.*, **9**, 181–198.

R. M. Wilson (1974b), Non-isomorphic Steiner triple systems, *Math. Z.*, **135**, 303–313.

E. Witt (1938a), Die 5-fach transitiven Gruppen von Mathieu, *Abh. Math. Sem. Hamburg*, **12**, 256–264.

E. Witt (1938b), Über Steinersche Systeme, *Abh. Math. Sem. Hamburg*, **12**, 265–275.

Index

The pages on which definitions occur are given in **bold** type.

C

O

P

Q

T

U

V

W

X

Y

Z

List of Authors

This list includes the names of all authors cited, and the chapters where quotations or references occur.

Printed in the United States
By Bookmasters